On To Mars!

Vladimir Pletser

On To Mars!

Chronicles of Martian Simulations

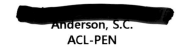

Anderson, S.C.
ACL-PEN

 Springer

Vladimir Pletser
Technology and Engineering Centre
   for Space Utilization
Chinese Academy of Sciences
Beijing
China

ISBN 978-981-10-7029-7   ISBN 978-981-10-7030-3   (eBook)
https://doi.org/10.1007/978-981-10-7030-3

Library of Congress Control Number: 2017958555

Printed on acid-free paper

This Springer imprint is published by Springer Nature
The registered company is Springer Nature Singapore Pte Ltd.
The registered company address is: 152 Beach Road, #21-01/04 Gateway East, Singapore 189721, Singapore

To Dimitri, Matt, Jack, Anton, Ben, Tom,
Eloïse, Lara, Romain, Victoria, Tigran,
Niels, Lea,
the next generation
with the hope that they will follow the
Martian way.

Acknowledgements

To many people, of course, for having conducted these simulation campaigns, such a great adventure.

To my family first, Jayne and Dimitri, for their support and patience.

To my friends for their help, and they are too many to mention them all here. Nevertheless, a special mention to Véronique, Philippe, Michel, Christophe and Bernard for their help to prepare these experiments; Christian, Théo and Juan for their help to circulate the information; Robert, Bill, Kathy, Steve, Charles, Andréa, David, Jan, Nancy, Euan, Anouk, Danielle, Stefan and Jeffrey my companions in adventure in the Arctic and in the desert; Aziz for his hospitality; Colleen, Pascal, John, Joe and DG Lusko for their logistical help in the field.

To the *Institut de Physique du Globe de Paris* for the loan of the geophysical instrumentation.

To the European Space Agency for the loans of the photo cameras and the computers.

To Michelle who typed parts of the manuscript. To Charles, Jayne, Dimitri, Perin, Brigitte, Qian and Yang for discussions and corrections at various stages of the English manuscript.

To Yang, Qian and Jian for making this book possible.

And then, to Konstantin, for his appeal towards space and for his inspiration to many for the exploration of other worlds.

The Author is supported by the Chinese Academy of Sciences Visiting Professorship for Senior International Scientists (Grant No. 2016VMA042).

Acknowledgments

Contents

Part I   The Arctic

1   The Arctic Before   3

**2   The Arctic During—Flashline Mars Arctic Research
Station Diary**   9

**3   The Arctic After—What Have We Learned
from This Simulation?**   65

Part II   The Desert

4   The Desert Before   75

5   The Desert During—Mars Desert Research Station Diary   79

6   The Desert After—The Last Day and the Return   123

7   What Did We Learn from This New Simulation?   129

Part III   The Desert Reload

8   The Desert Reload—Before   135

9   The Desert Reload—During   143

10   The Desert Reload—After   181

Part IV   Mars Tomorrow

11   Mars Tomorrow   209

References   241

Some Web Sites To Know More About   251

Introduction

Hello!

My name is Vladimir Pletser. As my name does not indicate, I am from Belgium but I feel more European than anything else. I would like to let you know in the following pages about my passion for space and for planetary exploration. Space, astronautics and the exploration of new planets have occupied my imagination, my dreams and my thoughts since childhood. In all logic, my studies and my career were oriented towards space research.

I have collected several degrees. In chronological order, an engineer degree in mechanics with a specialization in systems and dynamics, a special master's degree in physics with a specialization in space geophysics and a Ph.D. or doctorate in physical sciences, with a specialization in astronomy and astrophysics. But, don't worry, I am not a thick head. I have done a lot of sports and still do. And I was always looking for adventure, either related to space or sport.

My space career really started in 1985 when I joined the European Space Agency, better known under its acronym ESA, sometimes referred to as the European NASA. In fact, I worked for 30 years as Senior Physicist–Engineer (senior, not senile) in the Microgravity Projects Division at the European Space Research and Technology Centre (or ESTEC), located in Noordwijk in the Netherlands.

I was in charge of two programmes. First, the development of instrumentation for physical research in weightlessness (also called microgravity). I have participated in several space missions from the ground on Spacelab, the Mir space station and the International Space Station. Second, I was responsible for the ESA parabolic flights, those special flights made onboard laboratory aircraft that follow trajectories shaped as parabolas or camel humps and during which a weightlessness environment is created for about twenty seconds. I have performed so far more than 7300 parabolas, which is equivalent to 39 hours and 18 minutes of accumulated weightlessness time or slightly more than 26 earth orbits in slices of twenty seconds.

Since 2016, I am working as Visiting Professor–Scientific Adviser at the Technology and Engineering Centre for Space Utilization of the Chinese Academy of Sciences in Beijing, China, to support microgravity projects for aircraft parabolic flights and the future Chinese Space Station.

I was lucky enough also to have participated in astronaut candidate selection in Belgium in 1991. I was selected as Laboratory Specialist candidate with three other candidates and a Pilot candidate.

Space is something that I know a little bit about. But I have not yet gone into space, although it nearly happened a few years ago. But this is another story that I will tell you some other time.

I was fortunate enough to have had the honour of being selected to take part in two international campaigns of simulation of Mars manned missions, the first one in 2001 in the Canadian Great North, beyond the polar circle; the second one during the Spring of 2002 in the desert of Utah, USA. I went back in the Utah Desert for the third simulation in 2009 as Crew Commander for a programme called the EuroGeoMars that I co-organized with a colleague at ESTEC.

I kept each time a diary where I noted my thoughts, the experiments that we were conducting during extra-vehicular expeditions, the crew activities and everyday life little events of these three campaigns. It is these diaries that I want to share with you in these pages, with some comments and explanations.

But why willing to go to Mars? Let's first meet this little sister of planet Earth.

Mars is not far. Counting from the Sun, it is the fourth planet of our Solar System, the next-door neighbour of our Earth. Mars is not far on an astronomical scale. Let's look at some figures. The Earth revolves around the Sun at a distance of roughly 150 million of kilometres (about 93.2 million of miles), a very small distance at the scale of the Universe or even at the scale of the Solar System. Nevertheless, it still takes eight minutes for light to travel this small distance at a speed of about three hundred thousand kilometres per second (about 186 thousand miles per second). The light that left the Sun when you opened this book arrives to you only now (well, it depends if you are a fast or a slow reader, but let's not digress). This average distance of 150 million of kilometres (93.2 million of miles), let's take it as a unit and let's call it the astronomical unit, like Astronomers do.

Mars also revolves around the Sun, in the same direction, but a little farther, only half a time farther than the Earth on average, at one and a half Astronomical units. Mars is then closer to us at certain moments than the Sun. Mars is in fact one of the four planets called telluric, i.e. that are similar to the Earth. The other two are Mercury (very close to the Sun, it is by far too warm for Mercury to have an atmosphere) and Venus (second planet from the Sun, very warm also, more than 400 degrees centigrade (about 750 °F) on average because of the greenhouse effect).

To put things in perspective, here are some comparison values between Earth and Mars.

Mars is smaller than the Earth and your weight on Mars would be less. The diameter of Mars is a little less than 6800 km (4200 mi), while the diameter of Earth is a little less than 12800 km (7900 mi). The total mass of Mars is a little more than a tenth of the Earth's mass. The average gravity on the surface of Mars is 38 percent of Earth's gravity, that is, if your weight on Earth were 70 kg (154 lbs), it would be only 26.6 kg (58.6 lbs) on Mars.

The Sun also rises and sets every day on Mars. A Martian day is slightly longer than an Earth day, only 40 minutes longer. On the other hand, the Martian year is longer than on Earth, with a value of 686 earth days, or nearly two earth years.

It is generally colder on Mars. The average temperature on Earth is about +14°C (about +57 °F), while on Mars it is about −53°C (−63 °F) on average; but like on Earth, temperature changes between night and day and varies from the equator to the poles. On Mars, the equatorial temperature can vary from +22 °C (+72 °F) at noon to −80 °C (−112 °F) at night, and temperature at the poles can fall down to −120 °C (−184 °F).

Earth has an atmosphere of oxygen and nitrogen at an average pressure of one bar (14.5 psi). Mars also possesses an atmosphere, but made of mainly carbon dioxide (or CO_2, one atom of carbon to which two atoms of oxygen are attached, the gas that we breathe out, therefore, absolutely unbreathable) at an average pressure of 7 to 8 millibars (about 100 to 115 milli-psi), or about 120 to 140 times less pressure than on Earth.

The Earth has a magnetic field that protects us from solar and cosmic radiations. Mars also possesses a magnetic field, but extremely weak, half a million times weaker than the Earth's, and scientists ask themselves if this field is simply a remnant of a past geological activity or if it is still driven by a liquid core.

These are some interesting comparison facts that show that Mars is similar to our Earth by some aspects and very different on other points.

But Mars is still our neighbour. And as mankind is in good terms with its neighbours, it seems logical to go and visit our sister planet from time to time. And this is what mankind has been doing since more than 50 years ago, as our first spacecraft to visit Mars was launched in November 1964 and flew by in July 1965.

So why this interest for our sister planet? Why are we absolutely willing to go? Why would we want to send a manned mission? These are simple but difficult questions at the same time and answers are multiple, and so different that each one of us can find his or her own answer.

Let's start with the simple answer. Since the beginning of times, humans have dreamed of flying and leaving the surface of the earth. In ancient Greece, Icarus tried by glueing wings on his arms with wax. Bad mistake: according to the legend, he would have gone too high and the heat of the Sun would have melted the wax, making him fall in the sea that bears his name since. Let's jump a few centuries and many other legendary attempts, some serious, some humoristic, to come to the dawn of the twentieth century. A Russian scholar, named Tsiolkovsky, having written several papers on space travels, had these profound words: 'The Earth is the cradle of mankind. But nobody stays for long in the cradle!'

And this is very true. It is the same spirit, the thirst for discovery, the same need of adventures and dreams, but also sometimes economical and political realities, that have pushed Christopher Columbus to sail West, Charles Lindbergh to cross the Atlantic on a heavier-than-air craft, Werner Von Braun to build rockets to leave the earth surface, the US astronauts and Russian cosmonauts to conquer space around the Earth and the Moon. There is a long list of dreamers that found many reasons to go farther, higher, to see what is out there. It is this innate need in the human being that pushes him or her to surpass him- or herself and to discover new horizons and new spaces. And here is the first answer, so simple: Why would one go to Mars? Simply because Mars is there, and nobody went yet, besides through automated probes and instruments.

The second answer: Well, scientists, and I am one, will answer to explore, to try to understand the similarities and differences with our planet, to try to understand how another planet is made, in order to better understand our own planet. And this is also a good reason. Because sometimes science makes giant leaps forward, thanks to ambitious research and exploration programmes. But do these programmes need to be manned? Is it really necessary to send men and women to other planets to study them? Can't we do better and cheaper with good telescopes and automated missions like we have done so far? Well, this is an old debate: humans or machines. A machine is cheaper, it does not need food, warmth, a pressurized and breathable environment, and to be brought back eventually to Earth. Yes, but proponents of manned space flights would tell you that a machine would do only what it is built for and programmed to do and nothing else. A machine would never be as intelligent as a human being who can adapt him- or herself to new or unforeseen situations and make decisions on the spot, *in situ*. Yes, it is an old debate, and nobody is totally right or totally wrong. In fact, the two are complementary: some tasks are perfectly and better conducted by machines, and other tasks cannot be performed by machines. The intelligence that built these machines must also be present on the spot. So, the second answer: why go to Mars? For science and for exploration.

But there is more to it, much more. Mars is a new territory, a new planet, uninhabited. At least, that is what is thought today. If there would be life on Mars, it would not be little green men (we would know already if they were there), but maybe some form of bacteria or some other primitive life form. I'll come back to that question. No, what I mean is that Mars is a place where the human being can explore a new environment, study it and, while respecting it, eventually settle and start new colonies. That is right, like the first European colonists discovered the American continent and settled there. But, contrary to what happened a few centuries ago on Earth, mankind must learn the lessons of history and respect this new environment and any life forms that might be discovered.

But we are not yet there. This is a long-term goal. We would still need several decades, if not several centuries, before seeing the first human colony on Mars. But here is the third answer to the question 'why go to Mars?' To settle there, to start human colonies on a new planet and to eventually swarm the Solar System and much, much later the rest of the Galaxy. But these are very ambitious and very long-term goals.

Other reasons: I will let you think about it. You will find at least one, personal or not, for or against, positive or negative and optimistic or pessimistic. And related to these reasons, other questions, political, philosophical and economic. For example, has mankind the right to appropriate a new planet, to occupy it, to exploit it? Following which modalities, which rights and duties, in which order? What should mankind do if a primitive life is discovered on Mars? Here are, amongst others, the questions that are debated by the proponents and the opponents of space exploration during international conferences and meetings. Practically, how should we do it? By answering to the call issued at the end of the eighties by the US President George Bush for a manned planetary exploration programme, NASA has proposed an impressive plan of building a huge spaceship that would take a human crew to Mars with all that would be needed for the back and forth journey and for the stay on Mars. This plan would cost trillions of US dollars, a colossal amount with an impressive number of zeros. Perfectly unrealistic and impossible to finance. In reaction, a group of engineers and scientists, some working at NASA at that time, have created *The Mars Society*, and proposed a counter-plan that I will not detail here. It would take too long and it is not the purpose of this book. I invite those of you interested to know more about it to read the books written by Robert Zubrin, the founder of *The Mars Society*, that explains very well why the NASA approach is unfeasible. On the other hand, the counter-plan proposed by Dr. Zubrin and by *The Mars Society* is based on the principle of 'travel light and live from the land'. In summary, the idea is to send first an automated ship that would land on Mars and that would be the return ship of a future crew. An automated chemical plant on board this first ship would start to produce methane by chemical reactions of the carbon dioxide of the Martian atmosphere (don't worry, I will not bore you with details). Then, a crew of four, five or six persons would set out in a spaceship, a crossing between an Apollo capsule and a habitable module, that would land close to the first ship. The first leg of the journey would last about 6 months, the crew would stay on Mars about a year or a little bit more, and would come back with the first ship for the return leg of again about 6 months. Then, the second crew could be sent and so on. Further details can be found in Dr. Zubrin's books in the reference list.

All this to tell you that a manned mission to Mars is already feasible today, with the technology and knowledge that we have presently. Of course, there are still some details to settle that are discussed in the last part of this book, but Mars is actually within the reach of our technology. We know much more about Mars today than we did about the Moon when the first US astronauts landed in 1969, less than 10 years after the call of US President John F. Kennedy.

But let's come back to *The Mars Society*. This Martian society is an international organization, composed of passionate people, including scientists, engineers, but also people who are simply passionate of space and planetary exploration. *The Mars Society* is a private organization, functioning on private funds, without governmental support. *The Mars Society* was initially installed in the USA, but several chapters exist now throughout the world, in Canada, Australia, Japan, Europe, France, Germany, Great Britain, the Netherlands, Belgium, etc. Anybody can become a member; it would cost you only a few dollars, or Euros or Pounds to become a member. The goal of *The Mars Society* is to promote the idea of exploratory manned missions of the planet Mars with three types of actions: first, education and awareness actions towards the general public through conferences, public events, debates, etc.; second, lobbying actions directed at international space agencies and political authorities; and third, more technical actions to demonstrate the feasibility of this kind of missions and to research practical, operational and psychological aspects of a Martian crew living together for several years. For these purposes, *The Mars Society* has established a Martian Habitat (or Hab for short), a sort of module used as a manned base, installed on earth in a remote, difficult to access location answering several extreme environmental conditions, what is called a Martian analogue on earth, i.e. a location on earth that resembles by certain aspects what can be expected to be found on Mars, such as climatic, geological and eventually biological conditions.

I had the chance to take part in three such campaigns of simulation of Martian missions. I have learnt a lot from them and I like to think that I have contributed a little to the preparation of a future first manned mission to Mars in a few years.

There is another major reason for mankind to start a journey to our neighbour planet, probably the most important reason. And it is simply the dream. Yes, the need to dream and imagine. The dream of the child that is still within us, at any age.

I invite you to share this dream, simply.

Vladimir Pletser

Part I
The Arctic

Chapter 1
The Arctic Before

One day of December 2000, I ran across an article in *Space News* on Robert Zubrin and *The Mars Society* telling about the installation of the first Martian module on Devon Island, an uninhabited island of the Canadian Great North, way beyond the Arctic Circle. This article announced also the beginning of a series of international campaigns of Martian mission simulations that would take place in a habitat similar to a habitable base that would be landed on the planet Mars in a few years. *The Mars Society* organized during the summer 2001 a first international campaign of a Mars mission simulation. The goal of this simulation was to show the technical, scientific and operational feasibility of a manned Martian mission. *The Mars Society* was looking for volunteers to take part in these campaigns.

I remember also the article title: *Volunteers needed, no pay, eternal glory.* I liked this title and I got enthusiastic about the described activities. I visited the web site and it finished convincing me. Here was something I had to try.

The announcement on the web site gave a few more details. Six crews of six persons each would occupy this habitat in turn during rotations of ten days in order to conduct scientific experiments in geology, geophysics, biology and group psychology. The emphasis was put on the operational aspect of this simulation as the Martionauts would move around in simulated space suits outside the habitat, they would exit a decompression airlock following a strict protocol, and they would communicate by radio with a control centre with a 30-min delay like for a real Mars mission.

You need to be in good health and to have some experience of wilderness and field operations, as far as possible in extreme environments, which I have gained through my professional activities in zero gravity during parabolic flights and through my private hobbies (I am a confirmed diver, I have a private pilot licence and I did some parachuting).

You needed to have an experience of living and working in a multicultural and international environment, and you needed to speak English: no problem, as I work at ESA which is an Agency gathering together 15 different nationalities and whose working tongue is English.

© Springer Nature Singapore Pte Ltd. 2018
V. Pletser, *On To Mars!*, https://doi.org/10.1007/978-981-10-7030-3_1

You needed to have some experience in communication to relate and explain the goals and activities of this kind of simulation: again no problem here, I regularly publish scientific and technical papers and I regularly give conferences on space and space research to the general public and in schools.

You needed preferably to propose an experiment to be conducted during the simulation: also feasible, but I needed to think about that one.

And finally, you needed to send in three names of persons who could attest of the qualities of the applicant. I immediately thought of Professor Paul Pâquet, Director of the Royal Observatory of Belgium and the promoter of my Ph.D. thesis; Dr. Véronique Dehant with whom I finished my Master degree in geophysics and who is working at the Observatory as scientific responsible for automated experiments that would fly to Mars in 2007 on the Franco-American mission Netlander; and finally my friend Jean-Francois Clervoy, ESA astronaut, and with whom we started the parabolic flight programmes on the CNES Caravelle at the end of the eighties.

Talking with Véronique about ideas of experiments to be performed by a first human crew on Mars, she first suggested dusting off the solar panels that would be deployed on Mars. It is true that solar panels on Mars would suffer a lot from dust storms that can last several days or weeks and that astronauts would have to clean them off from time to time. But I told her that I already share the housework at home and that it reminded me too much of the story of the monkey and the astronaut. During one of the first space missions, a monkey and an astronaut are sent together in space with separate instructions contained in sealed envelopes. Once in orbit, the monkey opens his envelope, reads the instructions, and starts to push buttons, to switch on flashing equipment, to correct the trajectory, to insert samples in instruments, and to do a lot of seemingly complicated things. The astronaut, very impressed, has a look at his watch and sees it is time to open his envelope. He reads: open the fridge, take banana, peel banana, give banana to monkey…

Right. We needed something a little more evolved than this, using the human being in the scientific decision process. Véronique put me in contact with one of her colleagues, Dr. Philippe Lognonné of the Institute of Geophysics of the University of Paris VI, and who is also responsible for a geophysics experiment in seismology that will fly on the Netlander mission. He immediately proposed an interesting idea that seemed excellent to all of us: to conduct a seismologic experiment to detect underground water pockets on Mars.

I sent an application letter with the required information and this experiment proposal.

I have learnt much later, while talking with representatives of *The Mars Society* who I have met, that 250 scientists, engineers, and space experts throughout the world had answered this call and that the selection was difficult, intense, and very strict. Only ten candidates were retained, including three Europeans. I was selected to participate in the second rotation, from the 8th till the 17th of July 2001. And our Franco-Belgian geophysics experiment had been accepted by the scientific committee of *The Mars Society*.

Everything was thus going well. But, everything remained to be done and the rest still had to be prepared. Again, I contacted Philippe Lognonné and he kindly offered to provide the necessary equipment for this experiment. Furthermore, he proposed also to give me the specific field training to perform this experiment and he invited me to spend two days at the Geophysical Centre of Garchy, in the middle of France, close to the Loire River.

In parallel, I needed also to prepare adequately all my personal equipment for a polar expedition. Dr. Pascal Lee, from the NASA Ames Centre and who is also one of the officers of *The Mars Society*, sent a long triple list of mandatory, highly recommended and nice-to-have material to take to the Arctic. For example: a polar sleeping bag (mandatory! Ah, yes, we would sleep outside in this polar desert), a warm waterproof jacket, hiking boots, transport bags, etc. We were advised also to wrap up all our belongings individually in small zip plastic bags; I did not understand immediately why, but I soon understood. Try to keep all your clothes dry in a bag exposed to snow and rain in polar latitudes and you will see.

But let us come back to the preparation of our experiment. We arranged to meet with Philippe Lognonné for a weekend at the beginning of June 2001. He came to pick me up at the train station at Pouilly on Sunday afternoon. It was rather warm and we went to sit on a terrace to sip some chilled white wine while discussing the exploration of Mars, the Netlander mission, and our experiment: fascinating. Then, we went on by road to Garchy, a few kilometres away, to spend the night at the Geophysical Research Centre. Before going to bed, Philippe gave me three syllabi respectively, one, two, and three centimetres thick saying "There you are, read this before tomorrow. You will have a lecture on it tomorrow morning". As it was already passed midnight, I confess to having only glanced through them, and although the subjects were very interesting, I probably fell asleep on them. The next morning, I met Dr. Michel Diament, a colleague of Philippe, who teaches these subjects to students in Paris and who gives them the practical field training. Michel led me through a revision of the theory of seismic wave propagation in soils, the refracted waves being those that would be of interest to us for our experiment. Luckily, I had read most of it the night before, and it allowed us to go quickly through the theory and to spend more time on the practical aspects of the experiment. While walking in this magnificent Research Centre, I learnt that it would be closed soon in the coming years for reasons of budgetary cuts and restructuring. What a pity! Such a place, so rich in research and lecturing possibilities, that can accommodate so many students and researchers from all over the world! At lunch time, I met so many people, lecturers, technical and administrative personnel, who wanted to meet "the astronaut who was going to Mars" and who came up to shake hands. What a welcome, warm and simple at the same time. I keep excellent memories of this day. But let us come back to the practice field. Michel showed me what I should and should not do while unrolling the several hundreds of meters of electrical and data collecting lines; how to configure the field computer that would acquire all the data; how to perform the first analysis on this computer; and also, how to install the seismic sensors and how to generate the seismic quake, either by a good shot with a heavy hammer, or with a geophysical gun. We had envisioned also using explosives but the laws and regulations in this part of Canada were extremely strict on the importation and use of explosives. We would then go with the hammer and the geophysical gun. We agreed with Philippe and Michel: I would write the procedures of setting-up, running, and disassembling the experiment and Michel would correct them if needed. An excellent day of work as I like them: a dense learning session under the sun. Philippe brought me back to Paris to catch a TGV in the evening to go back to Holland. I had to be back in the office the next morning. It was a rather busy professional period as I was responsible for the development of an instrument to study the crystallization of proteins that would fly on the International Space Station in 2004 and that we would have to deliver in mid-2003 to be ready in time for the launch. We were in the middle of a formal project review and e-mails and faxes were piling up on my desk. I still had to complete the list of equipment to take with me and I spent all the rest of the afternoon running to all the sport and expedition shops to find the various items that were still missing.

A few days later, I learnt that Robert Zubrin and Chris McKay, another big name at NASA for Mars exploration, would take part in a conference organized by the Dutch chapter of *The Mars Society* at the Delft University. A great opportunity to meet Robert Zubrin himself and to discuss our simulation, as he would be our crew Commander. An excellent speaker, Robert Zubrin captivates and convinces you that it is time to leave for Mars right now, immediately, or at least tomorrow at the latest. He manages to convince you by reason or by emotion and to convey the idea that a new world is there, waiting for you. After his talk, we spent a few moments together to discuss the future simulation and our experiment. He was enthusiastic and agreed to share the shipping costs of our equipment, three large crates with a total of 130 kg. Philippe Lognonné proposed to cover the shipping costs from Paris to Ottawa, Canada, and Robert accepted that *The Mars Society* would cover the remaining costs inside Canada, which are not negligible. If you look on a map, you will see that the trip from Ottawa to Devon Island is longer than the transatlantic leg of Paris to Ottawa. We discussed also the practicalities of the simulation campaign and I learnt that the Discovery Channel, a TV education channel very popular in England, Northern Europe, and the States, would be one of the sponsors of this first international campaign. They would be present on Devon to film all operations in order to prepare several one-hour shows for broadcast in the fall in the States and in Spring 2002 in Europe.

So, it looked like everything was falling into place for this campaign.

I still needed to find some media support in Belgium to report on the simulation campaign and on our Franco-Belgian experiment. I contacted my long-time friend Christian Dubrulle from the Belgian newspaper *Le Soir*, who flew with us several times on parabolic flights. He was interested by the idea and agreed to meet Véronique Dehant and myself in Brussels. That was an excellent idea as I think it is important to make the general public aware that researchers in Belgium and France are amongst the best in the world for certain types of geophysical investigations and that they have a unique experience in earth and Martian geodesy and seismology. It was a done deal: Christian agreed to publish a few articles on the days that we would conduct our experiment. I would send daily reports of our activities, as I would also do for the e-journal on the ESA website. I would go with two still cameras: a normal one and a digital one, on loan from the ESTEC photographic department, to illustrate our expedition.

So, after having run back and forth, all was nearly ready. But I realized a few days before leaving that I did not have the time to mentally and psychologically prepare myself for this isolation campaign. Would I be able to stay locked inside for ten days without going out except for being covered from head to toe by an extra-vehicular suit? What would happen if one of the crewmembers did not get along with the others? How would the team leader play his role? I had the opportunity to meet him and besides being a brilliant mind and an excellent orator, was he also a leader, a team captain? Could one really stay locked in for ten days while rationing water? After having given it a long thought, I assumed that yes: people have survived for much longer with less while crossing or getting lost in deserts. But here, we had to continue to work, function, and live within a team. In any case, I was already so far in the preparation of this campaign and I had already involved so many people that it was too late to turn back: it had to work and to be a success. These few moments of doubts went away rapidly and I said to myself, as I usually do: after all, let's go, we'll see. Everything was there, everything was ready for it to work, and there was no reason to stop. Nevertheless, I was happy after all to have pondered over these questions. It showed that I was still lucid and aware of the difficulties that this endeavour could bring. I still had some days to prepare myself mentally to become a Martionaut leaving Earth civilization to go and conquer simulated Mars in the Arctic for a few weeks under the midnight sun. A few days and I would be on Mars in the Arctic.

Chapter 2
The Arctic During—Flashline Mars Arctic Research Station Diary

Note

This is the daily diary that I kept during my travels and during my stay on Devon Island. Large excerpts were published on the ESA website and in the Belgian newspaper '*Le Soir*'. Here it is in its entirety. Dates are indicated and day 0 corresponds to the planned day of entrance in the Mars Habitat.

Tuesday 3 July 2001, day-5

First day of the trip. A very long day. From Amsterdam to London, and then to Edmonton by AirCanada. Excellent flight. I had some salmon with a Canadian beer. Superb taste, the beer reminded me of the purity and freshness of the Colorado Coors that I tasted for the first time some years ago, while on a Rugby tour with a club from Brussels. But the taste still has nothing to do with our Belgian beers.

I am writing these lines from my hotel room in Edmonton. I have just read the chapter of the "Rough Guide to Canada" about the Northern Territories and the Nunavut, and other chapters on the history and geography of Canada. Quite impressive and interesting. The funny part is that the only place that they describe as being the most accessible up North to trained people with special gear and a purpose is still nowhere close to where we go, as it stops at Baffin and Victoria Islands. We are going to an island even more to the North to a place called Resolute to catch the last plane that will eventually drop us at Devon Island. Resolute, if you look at a map of Canada, is the last Northern civilized dot on the map.

So, I should arrive in Resolute at 2 a.m., two in the morning the next day, local time, that is 7 p.m., seven in the evening, body time or European time. Not that I mind, I always enjoyed travelling long days chasing the sun and the light when flying westward. Actually, time will not be the essence during this trip and this simulation campaign, as first, up North, it will be midnight sun all the time, and second, with jet lags and no difference between days and nights, what is the point of counting the hours as we do in a normal environment? I believe that several of these

© Springer Nature Singapore Pte Ltd. 2018
V. Pletser, *On To Mars!*, https://doi.org/10.1007/978-981-10-7030-3_2

taken-for-granted references in our common lives under our latitudes will disappear during the next three weeks, to be replaced by new feelings and constraints.

For example, the next plane that I will take leaves tomorrow, Wednesday 4th of July, from Edmonton to Yellowknife to Resolute. Departure time is 21 h 05, arrival at 22 h 50 in Yellowknife and then departure at 00 h 15 to arrive at 03 h 36 in Resolute, flying by First Air, the AirCanada subsidiary for internal flights. Can you imagine under normal skies, take-offs and landings at times like these? Well, it seems that under polar skies, it is normal routine. As they say, when in Rome, do like the Romans do.

Well, I am nevertheless relatively tired today after all the stress and running of the last few weeks to finish the normal work at ESTEC, plus the additional work for the preparation of this expedition, and yesterday running around in Brussels to meet with some friends, among which journalists to talk about this trip. This morning I went back to ESTEC, to catch up on some e-mail and to get my specially coded calculator to access my lotus notes e-mail system inside the bastion of the ESTEC firewall. But OK, it was an interesting and long day as every time you embark for a transatlantic trip.

Right. Stop here, I want to read a little before sleeping. Bye for now.

Vladimir

Wednesday 4 July 2001, day-4

Second day of travelling. Well, here I am in the Great North. We have just landed in Yellowknife, capital of the province of the Northern Western Territories. I am sitting in this small airport, muffled up in my winter coat, in front of a big white polar bear, the artwork of a taxidermist.

The approach to Yellowknife by plane was really strange and beautiful. Lots of open spaces, lots of water, a huge bay filled with small and large rock islands, and with forests of pines and rocks on the main land. The time is 23 h and temperature is about 15 °C (about 60 °F). We had some rough turbulence on the way, which felt quite funny in this small old Boeing 727, but nobody was really bothered. Other passengers all looked like lumberjacks or Inuit Indians coming home. The first thing that strikes you when entering the airport is the number of people with Asian-looking faces, all of them Inuit. The signs are all in two languages, English and Inuit. Inuit sounds and reads strangely and exotically but very nicely in a certain way. The writing is somewhat similar to the old Runic of Lapps in Scandinavia. I suppose that there are some common roots.

Our flight to continue to Resolute is planned at 15 past midnight, but we were already told before leaving Edmonton that, due to bad weather in Resolute (fog apparently) and although the flight is still scheduled to go ahead and will attempt to land in Resolute, if it does not make it, it will return and come back to Yellowknife. Considering that, if the flight leaves at 00 h 15 and is scheduled to land at 03 h 35, if the pilots decide to turn back, we should be landing back at 05 h 30 in Yellowknife. Sounds nice to spend the night going back and forth to the Arctic Circle to eventually come back to the starting point...I don't think that a lot of people are looking forward to that. Especially, after being told that we will be on our own, as the company Air First will not assume liability or responsibility because of bad weather. Well, we'll see.

Otherwise, I had quite an interesting and relaxing day. This morning, after a big breakfast, I managed to get the computer connected and I could check my e-mails in Estec and send a few e-mails. In the afternoon, I went visiting the West Edmonton Mall. Many superlatives are needed to describe it: the biggest mall in the world, the biggest car park in the world, a scale one replica of the Santa-Maria ship of Christopher Columbus in a huge pool next to a dolphinarium, an ice ring where an ice hockey competition was being held, a swimming pool with wave machines and toboggans, a huge fair with roller-coasters (I went twice of course; it was great, fast but too short), three cinema complexes, a countless number of restaurants, bars, shops, and so on. I went to see 'Crocodile Dundee 3' (not as good as versions 1 and 2, but still good fun, with a new version of the 20-year-old hit by *Men at Work*).

Although I slept well last night, I still have moments when I feel tired and my eyes are closing. Well, I suppose it is normal as it is nearly twenty to midnight on Wednesday night, which corresponds to twenty to seven in the morning Thursday body time, and I still have a few hours of travelling to do. And still, light shines outside like it was the end of the afternoon or an early evening at home, not yet the midnight sun, but close.

I am attacked by swarms of mosquitoes even here in the airport and despite the air conditioning. That's something I forgot to take: mosquito repellent cream. I was told that they can be quite voracious at this latitude during the short summer. It is apparently one of the reasons why mooses and caribous migrate north when the weather gets warm.

They are calling for the flight. I'd better get going. See you later.

Vladimir

Thursday 5 July 2001, day-3

At last, Resolute! We flew in last night partly in the clouds (and it was rather bumpy), partly in blue sky. The scenery below was quite fantastic: frozen seas and lakes, dry ground and rocks, no trees, nor grass, little ponds like scars tearing up the terrain, all laying in the same direction. And eventually, coming over the southern frozen bay, we arrived in Resolute, home of 230 inhabitants and a few dogs, the second northernmost community. It is mentioned on the map as a relatively large dot, but when you see it from inside, well, it is different. We landed on a strip of rocks and pebbles, no real concrete or bitumen tarmac. The airport is basically one big room, where everybody meets and hugs. I met with several people, some of whom were flying on the same plane. Like often in small communities in remote places, everybody knows each other and everybody greets each other, talks to each other and wishes each other a good day. Which makes these people truly friendly because the polar day can last up to six months....

Resolute in the summer. *Credit* VP

I met Colleen Lenahan from NASA who provides logistic support to the NASA-HMP project. HMP stands for *Haughton Mars Project*, named after the Haughton crater where we will stay on the Island of Devon. It runs in parallel with the project of *The Mars Society*. I met Aziz Kheraj, a friendly "young" man of multiple origins. Coming from Tanzania, but of Indian origin, Aziz lived in Resolute for the past 20 or 25 years. Aziz, married to a charming Inuit Indian lady, the father of many and grandfather of even more, is an astonishing man. He runs the only acceptable hotel in Resolute and several other businesses in town. He is the man to talk to if you have any problem in Resolute: he knows everybody in the village and in the other surrounding villages (which means within a radius of several hundreds of kilometres…) and he always manages to solve the problems. Among the people on the same flight were Andy, a welder by trade coming from Newfoundland who fell in love with the Arctic 15 years ago (I rapidly understood why); Joe, an engineer who will work on a new power plant somewhere up North for an advanced weather station (no more details on this. He is usually quite talkative, but not on this subject; it is true that there are a few military men around and that this kind of station is not on any civilian maps…).

The weather is superb, blue sky and shining sun, and of course this is in the middle of the night as it is half past three in the morning, Canadian Central Time, one hour later than in Edmonton. We put the luggage in Aziz's van and off we go on a bumpy trail toward the village. No bitumen roads as well, only pebbles and stones. After a cup of soup at Aziz's hotel and some small talk, it is time to go to bed. I fell asleep several times in the plane but, although I feel tired, I do not feel like going to bed, as if a new energy was inside me, due most likely to the shining sun outside. I went for a walkabout around the village down to the bay, to touch the frozen sea ice. Paradoxically, the place is very dry: the air and the ground give an impression of dryness, contradicted by the sea ice a little farther. There is no vegetation in sight. Nothing, no grass, not a single bush, no birds, a bare desert of rocks, of pebbles. Oh yes! At last, I stumbled on two small little yellow flowers on the path. The few wooden houses are all built on small poles and you can see ice patches underneath them. It reminds me a little bit of the Svalbard community on the Spitsbergen Island, where I went a few years ago. Children play on their bikes with a dog. Astonishing, it is nearly four o'clock in the morning and the sun is high, nearly at the zenith. I arrived at the edge of the bay and the sea is not completely frozen, there is well over a meter from the edge to the ice besides a few floating ice blocks. I take a run up and hop! I land on the ice, of a strange blue-green colour. The people around here say that until a few years ago, the sea was completely frozen and that the ice was coming to the edge of the bay. Now the ice melts on the edge and the frozen sea recedes every year a little bit farther. Should we see this as a sign of global warming? Maybe…

In front of Resolute Bay. *Credit* VP

Eventually I came back to Aziz's place and went to bed. I slept a solid 6 h in a row. What a pleasure! A warm bed and uninterrupted sleep.

As I wake up, the weather has changed completely. The sky is grey and it is raining, it was even snowing while I was sleeping, but the snow disappeared under the rain. I missed the breakfast of course, but this is of no importance, as the lunch will be served in a few minutes. In these places where passing time is not given by the rhythm of the sun rising and setting, the only key moments allowing points of references in a period of 24 h (I hesitate to use the word "day") constantly sunlit are the meals, prepared and served at fixed times. If you are hungry outside these hours, there is always a pot full of warm chicken soup and a mountain of sandwiches renewed every day. The cook's name is Nick, an Englishman who spends six months per year in Resolute and the other six months in the Canadian West. I also met the other people staying in the hotel, waiting to leave for other polar destinations. There are some five or six persons from the *Discovery Channel.* Apparently, the entire *Discovery Channel* crew arrived twelve days ago and they got stuck in Resolute because of the bad weather, either here in Resolute or in Devon Island. The weather these last few weeks apparently was too rainy and it is very muddy on Devon Island, which renders the landing strip difficult to use. Part of the *Discovery* crew was transferred by plane to Devon Island two days ago, but the weather changed so quickly that the rest of the crew had to stay in Resolute. There is also another group of military personnel waiting to be transferred to this advanced weather station. There is not much to do in Resolute, except walking around and playing pool or watch TV and videotapes.

As mentioned earlier, Resolute is a dry place, i.e. without alcohol and beer. Which means that, after a few days, these people start to go in circles.

I got acquainted with Kathy Quinn, a young geologist from MIT (*Massachusetts Institute of Technology*) of Boston and also a Rugby player. We will be in the same rotation in the Mars Habitat. She arrived also yesterday but in the afternoon from Ottawa. We discussed the seismic experiment that we will be doing and she agreed to give me a hand, which is a relief for me, as I would need her expertise in geology and geophysics. Speaking about our experiment, Robert Zubrin called me this afternoon to say that the equipment was still stuck in customs in Ottawa and that some signature was still missing. So, I called Dr. Philippe Lognonné, the co-investigator of this experiment and who lent the equipment from the French 'Laboratoire de physique du Globe de Paris', to ask him to send immediately his signature to whoever needs it. So hopefully the equipment should be here in Resolute on Saturday. Later on in the afternoon, Colleen Lenahan announced that we will not be flown to Devon Island before Saturday or even Sunday morning, two or three days later than anticipated, due to the bad weather and to the fact that only one plane is operating, the other broke down some days ago. This island is definitely quite difficult to reach…

It is still raining, drizzling actually, and some patches of fog are coming and going with the wind. So, to fly this plane to Devon Island, you need to have sufficiently good weather both here and there; it is not enough to have a good visibility here if it is foggy over there and vice versa, since all polar flights are flying by sight only. Although we are still scheduled to enter the Mars Habitat on Sunday, we will have to adapt to a new schedule whenever we will be flown up there.

Luckily, Aziz's place is rather large and comfortable. Aziz has an Internet connection, and there are still a few videos that I haven't seen. In short, everything needed for people in transit who are blocked by the bad arctic weather. We spent the time with Joe taking pictures of a wolf and a musk ox (again taxidermist artworks) and of the bay. It is nearly suppertime and I start to feel really hungry, most likely due to the cold and rainy arctic weather.

So, signing off for today from Resolute, the last civilized place before Mars.

Vladimir

Friday 6 July 2001, day-2

Another strange day in the arctic. The 24-h light outside is totally disturbing. I woke up at seven this morning and was up the entire day, somehow expecting the evening but of course the evening never comes in the arctic summer. So although it is close to midnight on my watch and I feel tired, I don't feel like going to sleep, as it is bright outside like in the middle of the afternoon. So I adopt the local way of doing things, a little nap in the afternoon and in the evening and up and about the rest of the day or "night".

Talking with Aziz, he tells me that people around here live with the sun, that is that they spend long hours outside in the summer taking naps of a few minutes or hours here and there, while in the winter, during the polar night, they feel constantly tired and sleep most of the time 18 h per "day". I can easily imagine waking up in the "morning" in the middle of the night and thinking that, anyway, the sun hasn't risen yet and one may as well continue to sleep.

The weather today is still the same as yesterday, that is a "warm" 0 °C (32 °F) with cloudy skies and lots of wind from the North-North-West. Still not good enough to take off to Devon Island during the day. There was an attempt last night at 10 h 30 p.m. Sometimes, the decision is made within minutes when there is a break in the clouds. The pilot managed to land with some supplies for those who are already there, but apparently it was very close, so close that he decided not to fly back as originally scheduled at two a.m. So the *Discovery* half-crew had to spend another day here in Resolute going in circles in front of the TV and surfing on the web. Until finally this evening when on a short call, the weather cleared sufficiently on both sides for them to go practically without notice. One of the guys left apparently with only his socks on and his boots in his hands to be laced in the jeep on its way to the airport. So would we be able to leave tomorrow, Kathy and I? We'll see. Robert Zubrin arrives tomorrow and I am sure that this will help to move things up.

So what did I do today? Well, quite a few things. I went to do some shopping and I went for a ride with Joe to drive the Royal Canadian Army personnel to the airport as their plane was leaving as planned at noon.

With three Royal Canadian Army personnel. *Credit* VP

It is funny how you get quickly acquainted in circumstances like these, stuck in the same place for several days. People are also quite confident around here. No doors are locked and Aziz leaves his keys of the van telling us that we can use it whenever we want. It is true that thieves would have nowhere to go.

We went for a 2h walk along Resolute bay. It was rather warm at 0 °C (we are still in the middle of July after all), but the strong wind made it so cold. I had the feeling that my ears were going to freeze. But no, not yet. The stories of frozen limbs to be amputated would be for another time.

We worked also with Kathy. We studied the procedures of our seismic experiment and we talked about the importance of this experiment for the exploration of Mars: how we would find water of Mars.

Working on the seismic experiment procedures with Kathy Quinn. *Credit* VP

I will come back to that another time, but let's just say that finding liquid water on Mars is absolutely crucial if humans are to settle and live on this new planet.

I also read about the history of the place. So let's do a bit of history and geography. The Polar Circle is at a latitude of approximately 67° North. Resolute is located just below 75° latitude North, i.e. about 15 degrees of latitude or something like 1700 km from the geographical North Pole.

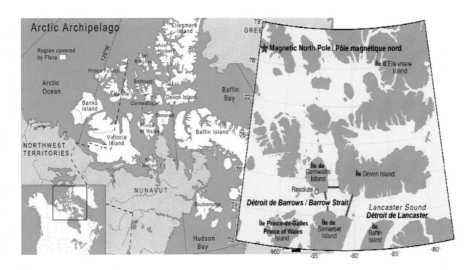

Relative positions of Resolute on Cornwallis Island and Devon Island. *Credit* Left: Canadian Museum of Nature/Musée canadien de la nature, Right: Bedford Institute of Oceanography/Institut océanographique de Bedford

If by any chance, you are lost on the surface of the planet and you do not know under which latitude you are, you can always rely on the old tricks of boy scouts: watch the stars, or look where the sun sets (if it sets…), or simply look at your compass. Well, technology being a wonderful thing, the comfort of modern life allows you to find an even easier way to find your latitude: just look at the parabolic antennas. As most of them are pointing towards TV or telecommunication satellites on the geostationary orbit in Earth's equatorial plane, they will be pointed nearly horizontally, which means that you are not far from the pole.

Joe in front of nearly horizontal parabolic antennas. *Credit* VP

Resolute is on Cornwallis Island in Nunavut territory. The name of the settlement in *Inuktitut*, the Inuit language, is *Qausuittuq*, which could be translated as "the place with no dawn". The English name of Resolute was given to the place after the HMS Resolute vessel sent by the British to search for the lost British expedition of Sir John Franklin that was looking for the Northwest passage to Asia in the 19th century. Nunavut, meaning "Our land" in *Inuktitut*, is also a new territory that was created in 1999 to recognize the importance of what is called here the "First Nations", that is mainly Indians and Inuits. Never say to an Inuit that he is an Eskimo, as this term is perceived as offending. It comes from the Algonquian language from South Canada and means "Eaters of raw meat". So voila, you know as much as I do about this place and its friendly people.

The only bond that links us to "home" is broken today, as the ESTEC web server is down. I could not therefore check my e-mails today and I do not know when and how I will be able to send this report. But hang on tight because it will come to you any way in the coming days. Joe showed me a website from which I could send my e-mails, but I would not be able to read my mails.

All in all, the morale is good, the soup is warm, and the bags are packed, ready to go at first call from the airport. But as much as I would like to be already on Devon Island, I don't mind sleeping another night in a comfortable bed in Aziz's hotel. Hopefully, the next report will be from Mars on Devon Island. So good luck to all of you and to Justin Hennin who's playing tomorrow Wimbledon's tennis final against Venus Williams (well, yes! Even in the arctic, we watch the news on TV). "*Allez Belgique*".

Vladimir

Saturday 7th July 2001, day-1

Well, finally things are shaping up and today it looks like we may be going. After having hung around for the n-th time in this beautiful place (Resolute), we received the word that the plane is up and running and that our turn will come sooner than we may think. The weather is absolute gorgeous and beautiful. The sun is shining and the temperature is still around 0 °C but facing the sun, you would think that it is actually hotter than July (like Stevie Wonder used to say). I went for a stroll again along the beach to admire the iced sea and I even walked on the water, which is easy when it is frozen. Then we made a few jokes with Kathy and Joe, like feeding the stuffed animals of Aziz.

In front of a stuffed polar bear in the hall of Aziz's hotel. Notice the size of the paws with respect to my head and I am standing! *Credit* VP

We did also some work on the computer. Always searching for one thing or another to do, these scientists… Scientists are like that, you have to forgive them.

In the meantime, Robert Zubrin, President of *The Mars Society* and our future Sim Team Lead (the commander of our crew rotation), arrived at last in the middle of the afternoon, still cheerful and full of energy as usual. As soon as he arrived, he started asking about shipment and flights and bookings and so on. Bill Clancey, a computer scientist involved in cognitive science research, who will also be on our shift, accompanied him. Bill is a big man, big in size, as he is tall and big by his writings as he is quite famous in his field.

We had our first team meeting to talk about the coming simulation. The situation has actually quite changed since our last contacts by e-mails before everybody left home. As the weather was so bad in the last days and weeks, the final preparation of the Hab, the Martian habitat, could not be completed in time. The first rotation crew, arrived more than two weeks ago, got stuck in the tent village of the base camp located a few kilometres from the Hab. This first crew under the leadership of Pascal Lee, a French Scientist working at NASA and whose family's origin is from Hong-Kong, could enter only yesterday in the Hab and their simulation period, initially foreseen for ten days like ours, was delayed by one week. Pascal was asking us to delay the beginning of our simulation by two days to allow them to conduct a minimum of their science program. Robert already granted them these two days and we were discussing the possibility of giving away an additional day. After some discussion about the impact on our experiments, we finally find a compromise for a half period of 24 h, meaning that we will be entering the habitat on Tuesday 10th July at 9 p.m. instead of Sunday 8th July in the morning.

Kathy and Robert walking toward Resolute's church. *Credit* VP

After dinner, we went to visit the office of the First Air Company to check for the geophysical equipment that was supposed to arrive from Paris via Ottawa. Aziz told us as well that his 16-year-old son, who was on Devon Island as a guide, was returning on the next flight as he had an accident. Nothing too serious, just the ankle superficially bruised by an ATV, one of these All Terrain Vehicles that are in such popular use here (nothing to do with the other ATV, the Automated Transfer Vehicle that delivered cargo and supplies to the International Space Station). So he was shuttled back by plane, accompanied by Dr. Rainer Efenhauser, of German origin and a Flight Surgeon at the NASA Johnson Space Centre, who was participating in the first simulation crew. He actually talked to me in German when I told him I was with ESA, while walking our way back through the village.

We were told as well that we might well be flying tonight at one a.m. or if not, on the next flight rotation at 4 a.m., but again these time indications lost their original meaning as the sun is up 24 h a day. It is a mere reference to your wristwatch, not to a particular moment in your day, as you do not have a "day" in the arctic. You sleep when you feel the need to sleep not because you are constrained by night hours.

Anyway, I thought that after all these days, and if we have to leave later on in the very early morning, I would be better off having a couple of hours of sleep. So off to bed I went and as soon as I fell asleep, Colleen, Robert, and Joe woke me up thumping on the door, to say that we were to go in 30 min, at 1 h 30 a.m. Just time enough to pack up the bags and to throw them in the van. Isn't it ironic? You spent three days lazing around with so much spare time on your hands that you do not know what to do with it, and the last second, you are told to rush off. But I don't mind as we are finally going to Devon Island, the location of our Martian environment on Earth.

Vladimir

Ready to board the Twin Otter aircraft to Devon Island. *Credit* VP

Sunday 8th July 2001, day-1 bis

The word of the day is "UNBELIEVABLE".

We jumped in the plane at 1 h 30 a.m. this morning after a few minutes of sleep and after having loaded ourselves, the luggage, boxes, crates, and so on in the plane. And off we went at two a.m. in bright sunshine. At last! What a liberation! We are going to Devon, Kathy, Robert, Bill, and I. The airplane is an old Twin Otter, which has ten seats and cargo space at the back of the cabin.

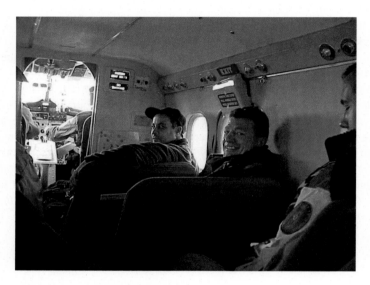

In the Twin Otter plane with Robert Zubrin. *Credit* VP

The flight itself is only 45 min, but it is like entering another world, as we leave one island for another, but different in its appearance, wilder, more tortured, chaotic, even more deserted. All water expanses are frozen up, except for some liquid water flowing between cracks. Not a patch of green, only rocks, snow, and ice. Nobody talks, everybody is mesmerized by this view of another world and anxious to arrive. The plane is losing speed and altitude in view of the landing. The light indication in the cabin reads "arrival on Mars in 10 min". Nice of the two pilots to put us immediately in the right ambiance. The pilots turn their face towards us through the open cockpit door and smile at us pointing at something on the ground.

And there it is! The Hab. Like a spacecraft landed on a pebble beach on the edge of a large circular structure: the ancient Haughton crater. Nothing is visible on the horizon in this barren world. The plane turns above the Hab to come in front of a bare strip of land and stones and slightly inclined. Yes, eventually, we can see some coloured points far away: it is the base camp made of a few dozens of tents.

The bare view on Devon Island, from left to right the mess tents, a rock formation, the Hab in the background and two more tents. *Credit* VP

The plane lands and continues to run on the up-hill strip, makes a half-turn and finally stops. We are welcomed by a dozen of persons, including John Schutt, the manager of the base camp, and Joe Amaralik, the head of the guides, who came from the base camp on their ATVs, the only mechanized moving vehicles on this island. We congratulate each other, we introduce each other, we get acquainted, we are here, we have arrived at last. It is three a.m. and it is warm, well it is the impression facing the sun, still high and bright. But the plane cannot wait and has to fly back. We unload the plane of all the luggage, boxes, crates, containers, and load in the plane the boxes, crates, and containers that need to go back to Resolute. Everybody helps and in a few minutes, everything is unloaded and reloaded. Before leaving for the base camp, I take a quick look around. Unbelievable, this Martian environment! Like if we were on Mars, only rough-hewn rocks all over with tormented bizarre shapes formed by the explosion of this huge meteorite 23 million years ago, and this futuristic Habitat only a few hundred metres away!

On Devon Island, from left to right, the Twin Otter plane, a rock formation and the Hab. *Credit* VP

But it is time to move. The luggage are quickly transported to the base camp on ATVs and trailers. We are walking to base camp by treading down a snow-covered rocky slope to a small river formed by the melting snow. We cross it with dry feet (thanks to Gore-Tex boots) and we climb up on the other side. John Schutt is waiting for us and shows us where we can set up our tents for the rest of the "night". I am so impressed by the scenery that I cannot take my eyes off the horizon and I keep looking around, comparing this rock with that one, and this snow patch with the next. From here, we cannot see the Hab, hidden behind a big rock. I was reading so much about this place on the website of *The Mars Society*, and here I am, in the middle of it. I know that it will go very fast, these ten days here and I want to take advantage of it as much as possible. I still cannot believe it.

The base camp is composed of three large tents, respectively the mess, a working tent, and the studio-tent of the *Discovery Channel* personnel. Two additional tents are a little bit further. These are for personnel usage to do…, well, let's say that the only furniture inside is a chair with a hole. A plastic bag is attached at the bottom of the chair, and once filled up, is removed, closed, and placed in another larger plastic bag. A simple code indicates that the place is momentarily busy: a red jerry can is placed in front of the tent door. Outside, two emptied fuel tanks with a funnel and a step allow gentlemen to…, well you see what I mean. During the first years, people used to do their businesses outside, in the wild. Later on, when the NASA-HMP people came back the following years, they noticed that vegetation was growing in places previously used, thanks to nitrates contained in human waste. It was then wisely decided to ship back everything to Resolute to preserve the pristine and wild state of this fantastic island. So, this is a part of the shipment that the plane brings back to Resolute at every rotation.

Further on, the tent village itself, where individual or two-person tents are mounted, strategically placed to avoid water streams. So, here am I, at 4 a.m. trying to set up in the Arctic a tent that someone lent me. Luckily, the spirit of mutual aid is still functioning, and after having helped my colleagues, they come in turn to give me a hand.

The tent village. Arctic camping at its best! *Credit* VP

Again, I could not sleep and I decided to wander off alone around camp. Wrong decision, as the next day we were told to avoid wandering alone outside the limits of the base camp because of possible encounters with polar bears. Well, luckily enough, nothing happened to me that morning, besides being once again totally mesmerized by the wild and majestic beauty of this place. What an unbelievable feeling to be in this out-of-this-world environment!

Well, eventually, I go back home (that is my freshly mounted tent), to download photos from the digital camera to the laptop. Then, it struck me: if someone would have told me twenty or even ten years ago that in 2001, I would be sitting in a tent in the middle of a Martian-like environment on an uninhabited Arctic island at 4 a.m. under a bright sunshine downloading digital images on to a portable connectionless laptop, I would have asked in which Sci-Fi movie he would have seen that. Yes, reality can quickly overtake fiction.

Anyway, after feeling tired of thinking so hard, I unroll my guaranteed −20 °C sleeping bag on a thin soil mattress. A little hard, but one has seen worse. Thus, everything is going extremely well. It is not even too cold. I fall asleep thinking that a manned expedition to Mars could happen also faster than one may think.

Remembering unconsciously that someone said something about breakfast at 7 h 30, I managed to wake up a few minutes before. Fortunately, we still have our watches; without them, we wouldn't know what time it is. The sun practically hasn't moved in the sky. I walk to the mess-tent, thinking that washing and tooth brushing could wait until I would be more awake and a little bit warmer. It is already crowded in the mess tent, the only heated place on the island. I find familiar faces of the *Discovery Channel* team to whom I didn't have time to say goodbye before they left a few days ago. Other people, scientists, engineers, and journalists are also there, trying to warm up. Somebody puts a plate in my hands with three freshly made and warm pancakes and a cup of tea. Mmmh! The simple pleasure of a warm breakfast. I introduce myself to the other members of the expedition. There is a group of engineers and scientists from the Carnegie Mellon University here to conduct a technology test of a sun powered robotic rover; several groups of journalists from *Discovery Channel* of USA and of Canada, from the *Popular Science* magazine, a group of biologists of the NASA Kennedy Space Centre, geologists and geophysicists from the group of the NASA-HMP, etc. In total, about 40 persons, three dogs, and a lot of portable laptop computers, cameras, radios, and … shotguns, these short guns popularised by the cowboy Jos Randall in the sixties. We discuss a few things, that is mainly the various experiments in progress, the exploration outings of the day in the crater and the expeditions of the geologists of the NASA-HMP group. All this at breakfast. No waste of time. Very well, I like that.

John Schutt takes us aside, the newly arrived group, and explains to us the base rules of the camp: don't wander off alone (hum!), always stay in group, how to use the toilets, that the shower is not yet installed but that it would be very soon, etc. We are instructed on how to use hand held radio transmitter-receivers ("remember the call sign is HMP7SFU"), on how to ride the ATVs (basically like a motorbike, except that it has four wheels instead of two; as I had a motorbike while living in Africa, it is not too much of a problem for me).

All-Terrain Vehicles are fun and easy to ride. They are also essential in this desert environment. *Credit* VP

And finally, on how and why to use shotguns. Well, yes, I know, this one is hard to swallow, I do not like firearms, but after having heard the reasons, well, I have to admit that I prefer to know that there is always one at hand. Polar bears are running along freely from island to island on the frozen sea and, in fact, we invited ourselves to their place without having asked for permission. If a bear is hungry, he may well be hunting and basically, for him, you are meat on feet. If he starts to charge you, he rushes at you at an average speed of 20 m/s, or 70 km/h (to recall, the world record of the 100 m is slightly less than 10 s, a mere 36 km/h). So, do not even contemplate running away! But, in addition, a bear is clever and he has a very developed sense of smell. If he smells you from afar, he will not come directly at you. He will prefer to follow you at a distance and turn around you far away to be against the wind, so that you cannot smell him (just as if we would have a sense of smell as developed as his…), waiting for the right moment to jump on you from behind a rock. Even more worrisome: a polar bear is big. Standing, he can measure more than two and half metres and weigh more than 500 kg. His paws are fit with rather impressive claws. A little blow with his paw, and there you are without a hand, or an arm, or even a leg. If by any (bad) luck, you are surprised by a bear, the best thing to do is to take off a piece of clothing and to throw it at him. Not that he is cold, but that would occupy him for enough time for you to grab the shotgun. So, if we are to go on an expedition or simply out of camp, we are supposed to have at least one shotgun within the group. Furthermore, no perfumes, or smelly aftershave and do not leave any food or anything that could have a smell whiffing far away. John shows us how to load, to aim, and to fire the shotgun (luckily) on a cardboard box. Well, I must say that I did not enjoy it too much. I am glad to have done it so I know how to use it, but I still think that the world would be a safer place without them, except for wild places like here where you are at the other end of the food chain.

John Schutt instructions on how to use a shotgun. *Credit* VP

Anyway, it is better to avoid at all cost to shoot at a wild bear. Even in case of self-defence, the killing of a bear is punishable of a hefty fine by the laws of Nunavut. So, caution and suspicion. I would not go again wandering alone as I did before going to bed last night.

All this took us the entire morning. The lunch is served at noon sharp, and is more than welcome, as despite being sunny, it is rather chilly and calories are flying away quite fast. The choice is rice or rice with either chicken white sauce or chilli con carne, all of it coming from tin cans. These cans would be our lot for the days to come, but it is tasty, it is warm, and it is necessary.

The mess tent at lunch. The only heated place on Devon Island. *Credit* VP

The weather started to deteriorate in the afternoon, with a cloudy sky and some wind. As we live primarily outside and under tents, an additional layer of jumper is necessary, which brings the total of layers to six to be able to continue to function normally. In fact, it all boils down to: a tee-shirt, a rolled collar jumper, another jumper, again another jumper, a sleeveless jacket, the heavy duty jacket, and a bonnet for the head; for the bottom, a long-john, a jogging trouser, and a heavy trouser above everything; a pair of thin socks, another pair of heavy socks, and the Gore-Tex boots. This is how we live practically 24 h a day.

Later on this afternoon, we have planned with Kathy to open the boxes of equipment for our geophysics experiment and to go through the instrumentation. We will try to do a dry run maybe tomorrow before we actually enter the Hab on Tuesday night.

Checking the geophysics experiment equipment. *Credit* VP

So, signing off for today. A day full of learning and productive in emotions. Yours, a-little-bit-more-Martian-today-than-yesterday.

Vladimir

Monday 9th July 2001, day-1 ter

Second day of work on Devon Island. The weather turned to warm this morning, up to 2 °C above zero, rather warm indeed, but it is bearable thanks to the falling rain. I woke up early this morning after sleeping more than 6 h in slices of one or two hours. First, a sleeping bag on a mat in a single tent set up on rocks is not the most comfortable place you could dream of, and second, with the ambient cold, I found out that one has to go to the loo more often. So it was always the dilemma of either staying in the warmth and hold on or dressing up quickly and walking the 200 m in the windy rain to the outside pee station to get some relief. Luckily, we were told to take a plastic bottle for this kind of need. Unfortunately, it is filling up quite rapidly. Another reason for waking up early was to clean up today. That is right. It is Monday morning and time to use towelettes to "wash up". What a pleasure to contort in this small tent, to take off layers of clothing (we sleep of course half dressed), to wipe up inaccessible parts of the body with these humid towelettes (some are still frozen), and then quickly put back on the still warm clothes. The last shower was nearly two days ago, and although it is cold, you sweat from time to time when working or getting in the tent. But eventually, one gets used to it and I start to understand why the inhabitants of the Far North prefer to keep the layer of sebum or perspiration as natural protection against the cold. I decide also to go to the river to brush my teeth. Brrrr! It was cold and one must really want to do it. Try it at home. Take some ice, let it melt, and as soon as it melts, use it to brush your teeth. Good morning!

Well, we are not en route to Mars to discuss trivial matters but these are aspects the Mars crews would have to face as well, even if, for the moment, we are still doing arctic camping.

Talking about camping, no bears in view so far and luckily I am protected by this special sign saying that I don't taste good and that all of my bones have been replaced with Titanium, that my ESTEC colleague, Tammy Erickson, made for me. I am sure that any intelligent bear that can read will understand and go away.

With the "No bear allowed" sign made by my colleague Tammy Erickson. *Credit* VP

The first crew is still in the Hab and yesterday afternoon they had their first extra-vehicular activity, or EVA (pronounce "ee-vee-ay") in space jargon. The rain was so heavy this morning that their morning EVA was postponed till the afternoon. While their simulation (or sim for short) was going on, construction work around the Hab was continuing. We were asked to give a hand to lay a pipe from the Hab to the small river running a few hundred metres down. It took us 2 h, Kathy Quinn, Bill Clancey, John Schutt, and I to install 600 m of plastic piping in the Martian Hab environment. We finished in time for the noon lunch, a pleasant Inuit stew, called "Anaq", made of beef, vegetable, and a few other things that I prefer not to identify.

In the mess tent with the ESA flag. *Credit* VP

The weather turned better in the middle of the afternoon, with the sun coming up and clouds leaving most of the sky. It sounds silly to talk always about the weather like there was nothing else to report. In fact, in the arctic, the weather is one of the most important elements that would condition your behaviour and your decisions. For example, we planned last night, after having verified the equipment for our experiment, that we would do a dry run today in the *Von Braun planitia*, the flat plain in front of the base camp. However, this morning there was no point in trying to detect water under the surface while it was raining and this dry run would not have been dry anyway. Furthermore, the ATVs were needed to support the EVA of the first sim team and we would have needed them as well to carry around our 130 kg of equipment. So we will try this dry run again tomorrow. Another example is the weather report that Bill gets from a Canadian Station that allows him to download the satellite picture of the area. A cold front coming from the North Pole, not far from here, is forecast for tomorrow morning with negative temperatures. So it will be additional layers to put on to sleep tonight and to work tomorrow.

Some words about science. We had a very interesting seminar yesterday evening given in the mess tent by Dr. Gordon Osinski (Oz for close friends), from the University of New Brunswick, Canada, talking about "Impact Craters". As we are staying on the rim of the Haughton crater formed by a crashing meteorite 23 million years ago, it was most appropriate. This evening, Kathy Quinn will give a talk on "Icesat, observing ice sheet topography change from space", her thesis subject. This arctic research camp we are in is the most interesting place to have talks like these.

About forty persons, most of them scientists involved in life in space and Mars research, are actually staying in the camp. Another US scientist asked us this morning not to clean or sweep up the floors of the three main tents as he intended tomorrow to collect samples of dust accumulated over the last week to compare to samples collected a week ago, to assess the microbial contamination of human crews in new environments. He specified that he would not sweep up individual tents for control. It is true that the environment is very dusty, even if it is raining regularly. As soon as it dries, the dust accumulates on clothes, instrumentation, computers, and in tents.

A few words on the crew of the second sim rotation. We will be six entering the Hab tomorrow. The Sim Team Leader will be Robert Zubrin, engineer, founder of *The Mars Society*, and fervent advocate of manned Mars missions. Dr. Charles Cockell, a biologist from the *British Antarctic Survey*, is part of the first sim team and will stay on the second rotation as well. Steve Braham, a physicist from the Simon Fraser University of Vancouver, and specialist in communications will also stay on from the first to the second sim team. You already know Bill Clancey, a computer scientist from the NASA Ames Centre, and Kathy Quinn, a geologist from MIT, Boston, USA. We should be entering the Hab tomorrow evening at 21:00 and I am looking forward to that. We were already contacted by e-mail by a group of American psychologists who proposed us to fill out during the sim a questionnaire to assess some human factor in living and working at the base camp and later on in the Hab. So science work as a guinea pig and test subject is already *en route* and I look forward to start the science work as an experimenter as well. It will begin full speed tomorrow once in the Mars Habitat.

Vladimir

Tuesday 10th July 2001, day 0

Busy day today. Cold as well, as announced yesterday by the weather report, but we were fortunate to avoid the snow that fell further down South. The sun shines now in a nearly perfect blue sky with a temperature of +3 °C. This morning, I mustered up my courage and I decided to go for a wash in the little river downstream. Mmmh! It makes you feel alive and kicking, some fresh icy cold water on your face in the early morning arctic wind. After breakfast, we held a meeting with the members of the second rotation to discuss the meals and the food that we need to take in the Hab for the coming weeks. It is mainly non-perishable dried and canned food, not so much the salad, fruit, and fresh vegetable type. But I suppose that a Mars expedition would go for that as well and for longer.

I tried with the help of Patricia Garner, a young English lady engineer from the Simon Fraser University in Vancouver and an ice-hockey player, to set up again contact with the ESTEC host website, but unfortunately all contacts this morning were made apparently impossible by a very slow and jammed satellite connection. To aggravate the matter, my PC laptop crashed down in the attempt. Why? Mystery. We were warned that the environment was very dusty and very cold for computers. Now I am reduced to work with *Windows* in a reduced safe mode, I who a few years ago were only swearing by *McIntosh*, but, OK, it still works. Kathy proposes very kindly to use her laptop to transfer my files and to send them by e-mail through the satellite connection from the base camp.

Later on, as the weather looked like it would stay stable for the rest of the day, we decided to have a dress rehearsal of the geophysics experiment and to have at last the dry run. With Kathy Quinn and Robert Zubrin, we packed the three boxes of 130 kg in total on a trailer of an ATV and we left base camp the three of us to go to the Haynes Ridge, the plain in front of the Hab. The Hab is actually located on the rim of the Haughton crater and from it, you have a breath-taking view of the crater and the pale grey breccia that filled the crater after the meteorite impact 23 million years ago. The crater itself is a complicated circular structure with several circles, the largest having a diameter of about 20 km. The central circle visible from the Hab is about 2 km and the view is really magnificent and out of this world, with lots of sharp rocks with colours ranging from brownish to dark grey, and patches of white snow.

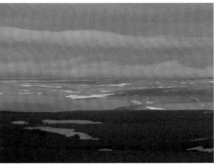

Two views of the Haughton crater from the ridge close to the Hab. *Credit* VP

Riding the ATV is quite an adventure itself. In order to preserve as much as possible the original and dramatic aspect of the landscape, several trails are already marked and we are requested to avoid leaving these trails. As the trail that we followed winds among the rocks, crosses small rivers, and patches of permanent snow, the second ATV got stuck in snow and nearly flipped over in the river on the way back.

We could perform our dry run of the geophysics experiment. It took us actually longer than anticipated, about two and a half hours for one run instead of the one and a half hour expected. Although we were not yet wearing EVA suits, the cold wind on this bare plain made it so chilly that we had to keep our gloves on and protect our ears and head. We managed eventually to set the line of 24 sensors perpendicular to the rim of the crater and to conduct the seismic tests using the sledgehammer. The idea of this test was to assess that everything was functioning properly. The instrumentation lent by the "*Institut de Physique du Globe de Paris*" (IPGP, Institute of Geophysics of Paris) did function flawlessly and some first measurements could be obtained on the underground structure of the crater rim.

Loading the geophysics equipment on ATV with Kathy Quinn. *Credit* VP

Performing the geophysics experiment in a dry run mode. Notice the Hab in the background. *Credit* VP

I feel much more relaxed now and I realize the importance of meticulously preparing an experiment like this. I owe a lot to Philippe Lognonné and Michel Diament of the IPGP who gave me an accelerated field training at the Geophysical Research Centre of Garchy a month before leaving. A first glance analysis could show already some asymmetry in the ground structure between the closest and farthest points from the crater rim, most likely due to larger compression of material close to the crater. This was an excellent first run that bodes well for the future experiments that we will be conducting during EVAs (Extra-Vehicular Activities) from the Hab in the coming week. As this test took us most of the day, the rest of the afternoon was spent preparing for entering the Hab and the hand-over between the two crews.

The sim team changeover is still foreseen at 21 h this evening. We already packed our bags from the individual tents and we are looking forward to sleeping this evening in the space ship.

The weather this evening is turning back to cold and cloudy with some fog. It makes it very humid and everybody is getting back to double layers of fleece and jacket.

So, my next report would be from the Martian Hab. While preparing mentally to be cut off from the rest of the world (except by e-mail) for a week, I'll sign off with a bold "On to Mars".

Vladimir

Wednesday 11th July 2001, day 1

First day spent in the Martian Hab. Great! We were able to sleep in the homey warmth of a space house. What a change after the arctic tents!

But before anything else, let me take you for a tour of our new home. The Hab is a cylindrical structure of 8 m in diameter and 6 m high, with two floors.

The Hab with the Martian flag on top and two ATV to the left. *Credit* VP

The ground floor has two entrances with airlocks used to simulate depressurisations and repressurisations of Extra-Vehicular Activities; a large room used as laboratory for experiment preparation and instrument storage; a small bathroom with a sink and a shower, and an incinerator toilet.

In the Hab, ground floor, left: the EVA preparation room with all suits hanging neatly; right: the lab area with the geophysics experiment equipment unpacked. *Credit* FMARS-2 crew

On the upper floor, accessible by a ladder against the wall, one finds the electronic working area along the circular wall with all computers, radios, and other electronic gadgets, the dining/working/meeting table in the middle, a small kitchen in the corner and six small bedrooms with a door: each has a small narrow stand-up space and a recess in a lateral wall that separates two rooms, one room having the recess on top, the other on the bottom.

In the Hab, second floor, left: top view of the living room with the central table and the circular table along the wall with all the computers and electronic gadgets; right: one the bedrooms with the recess on top for the sleeping bag. *Credit* FMARS-2 crew

There is just enough room to unroll your sleeping bag on a small bench. In the working area, three circular windows allow you to contemplate the Haughton crater on one side, the Haynes Ridge in the middle, and the lower canal, the small river running downstream.

View of the Haughton crater from the circular window of the Hab. *Credit* FMARS-2 crew

Everything is still freshly built and the previous crew had to finish up on some DYI work of painting, plumbing, and other chores. Well, we did our bit as well as this morning, the entire crew spent 2 h cleaning up the place to make it look like our home for the next week or so.

In this morning briefing, we decided on several rules on how our community will live for the next week. We agreed that one person per day would be in charge of preparing the food and cleaning up. We discussed the need to reduce the number of external visitors in the Hab to two persons at any one time, to make this simulation credible. *Discovery Channel* being one of the main sponsors of this campaign, they have a contractual right to have a cameraman inside at all times, night and day, and to film everything (well, within decency limits, of course). Bob, the cameraman, spent a few nights with us, sleeping in the attic, a small place above the six rooms, ready to film everything and anything, from the breakfast to how we brush our teeth. A little strange at the beginning, we get used to it rapidly and we do not see him anymore after a while. That is what he is asking for anyway.

The first briefing: from left to right, Bill Clancey, Charles Cockell, Vladimir Pletser, Robert Zubrin. *Credit* FMARS-2 crew

The first briefing in the Hab with the Discovery Channel team filming. *Credit* FMARS-2 crew

A first EVA will take place this afternoon, conducted by Robert Zubrin as EVA Commander, Kathy Quinn, and myself. It would be a walking EVA of 2 h in front of the Hab to search for fossils and other samples with biological implications. The EVA would be supported by the rest of the crew: Bill Clancey to document the technical aspect of donning and doffing the suits and of monitoring the communication exchanges; Steve Braham to follow the radio communications; and Charles

Cockell as the microbiologist expert to guide in the choice of rocks and samples to be picked up.

So, after a quick lunch that I have prepared (I took the first duty turn), we started at 13 h 30 to prepare for donning the suits. Although not pressurized, these are actually quite similar to real space suits. The suit is made of a heavy-duty material with a breast pocket and two other pockets on the legs. The suit goes over heavy warm clothes (after all, we are still in the Arctic) and is zipped at the back. A backpack, of about 15 kg, contains the simulated life-support systems, that is a water reserve to drink through a tube and a mouthpiece connected to the helmet, and a battery powered fan for air circulation through two tubes blowing in the helmet. The helmet allows you a nearly 180-degree visibility. You have to don a head set before fixing the helmet to the rest of the suit and to the backpack. Boots and gloves enclose the rest of the body. Two badges on each shoulder: one is the Martian flag designed by *The Mars Society*: blue, green, and red (I let you figure out why by yourself); the other is *The Mars Society* logo. Finally, a badge with our name is fixed by Velcro to the suit breast pocket.

Following the procedures to don the extra-vehicular activity suit. *Credit* FMARS-2 crew

It took the three of us nearly one and a half hours to don these suits. Not bad for a first time. Beating the previous crew who took 3 h for four persons to don the suits. It is true that once outside, we are not supposed to return because we forgot something. The simulation must be as close to reality as possible. Oh yes! I nearly forgot something important: there is no toilet on EVA, you have to take your precautions before.

The three of us enter the airlock, it is a bit tight but it just fits, and we wait the 5 min, simulating a 5-min decompression to equalize the inside and outside pressures. The outside door is opened at three o'clock sharp to find ourselves in front of our friend Bob and his camera. As soon as we are out, we realize that the radio communication does not function properly. Something is wrong with the Vox system that enables one-way communications as soon as you speak loud enough. However, this mode consumes more battery power and apparently, batteries do not recharge well in the cold and humidity of the Arctic. Nevertheless, continuing with hand signals, we decide to proceed as planned, as anyway we are staying close to and in view of the Hab. We start our rock and fossil collection. All of the rocks encountered date from Palaeozoic times, between 300 and 400 million years ago. Some are fossilized corals or shells, remains of intertidal seabed of a few hundred million years old. Others were thrown about by the impact that formed the crater, and are partially covered by a greenish grey foamy layer. We were later on told by Charles, our team biologist, that these are blue green algae, or more scientifically *Gloeocapsa sp.*

Our EVA lasted a little bit less than 2 h under the rain, and was relatively exhausting, as the bulky suit made each natural movement difficult to realize. We collected the sampled rocks with a long scooper and a metallic grabber so as not to oblige the Martionauts to bend too much forward. Because of the rain and the cold, condensation appeared inside the helmet, which made it even more difficult to see clearly, as the rain droplets were already covering the helmet external surface. To make matters even worse, I found that each time I leaned forward, water was escaping from the drinking tube in the helmet and flowed down inside my suit, making this EVA even more humid.

The extra-vehicular activity (EVA) with Kathy Quinn to collect fossilized samples. *Credit* FMARS-2 crew

Upon return, we went through the reverse procedure in the airlock and delivered the samples to the team biologist who started immediately to analyse them. The EVA was most interesting and instructive, but it was with pleasure that we removed our helmets and suits. During the EVA debriefing, it became clear that any field activity under EVA conditions would require more than double the normal time.

Pointing at the fossils. No, Robert Zubrin is in the background. *Credit* FMARS-2 crew

I then prepared the dinner, which consisted of cold tuna fish with olive oil, (warm) rice, and (warm) green beans, with syrup pears and apricots for dessert. Not too bad for a tin can meal!

Tomorrow would be another busy day, as two other EVAs are planned. The morning one would be a motorized EVA to check out the terrain a little farther away from the Hab, the afternoon one would be to conduct our geophysics experiment in Haynes Ridge, in front of the Hab.

As I still have to send my reports and as Kathy is still lending me very kindly her PC (my laptop is still working in a reduced mode), I'll finish here after a very exciting first day in our Martian Hab and I look forward to another field science day on Martian ground.

Vladimir

Thursday 12th July 2001, day 2

Second day in the Martian Hab. It was an excellent and very busy day. The journalist Frank Vizard, from the *Popular Science* magazine, was invited to spend the night in the Hab with us to follow a typical day of work in the Hab. We conducted a three-person EVA lasting 4 h to conduct the Franco-Belgian geophysics experiment. The goal of this experiment was to assess the feasibility of a seismic method by human operators to detect subsurface water on the planet Mars.

The presence of water on Mars is a subject of debate since long among scientists and the issue is far from being solved.[1] Water cannot exist in a liquid state on the surface of the planet due to the low atmospheric pressure (about 7–8 millibars, 125 times less than atmospheric pressure on Earth). Indeed, when the ambient pressure diminishes, water starts to boil and to evaporate at relatively low temperatures. Water can exist on the surface of Mars in the form of ice, as in the polar caps. Water in liquid form is hypothesized to be found underground, maybe trapped in water pockets in rocks. If this is the case (and most likely it should be), it is important to be able to locate these pockets, for two reasons. First, a human crew on Mars could try to tap the water in these pockets and use it for either their own consumption (drinking, washing, cooking...) or by dissociating it (a water molecule H_2O is made of an atom of Oxygen (O) to which two atoms of Hydrogen (H) are attached), to produce Hydrogen (to be used as fuel) and Oxygen (to be used as combustive to burn the fuel or to breathe). Second, if life is ever to be found on Mars, it is more than likely that it would be in the form of bacteria and not little green men like it is sometime believed. Bacteria were found on Earth in the most unthinkable environments, like at the bottom of the oceans, at several thousands of meters deep, under pressures several hundred times the atmospheric pressure, or in volcanoes at

---

[1]Since then, the Mars Odyssey spacecraft seems to have discovered in Spring 2002 indirect evidences of the presence of water ice under the planet surface, and it was later confirmed by the radar sounding of ESA's Mars Express mission.

very high temperatures, or in deep ice, at high altitudes above 70 km, etc. Bacteria can live like that in extreme conditions, without air, but not without water. Find the water on Mars, and you increase your chances of finding life.

This geophysics experiment was proposed as part of collaboration between scientists of the *Institut de Physique du Globe de Paris* (IPGP, Institute of Geophysics of Paris), the Royal Observatory of Belgium in Brussels, and myself. The experiment consists of deploying a line of 24 sensors (called geophones) to be firmly planted in the ground and connected to a data cable (called flute), itself connected to an acquisition system, a sort of field computer. A mini earthquake is generated artificially by hitting with a sledgehammer a metallic plate placed on the ground near a trigger geophone. The generated shock waves propagate in the ground in all directions, eventually reflect and refract on underground interface layers, between different kinds of underground materials. All signals, including those returned from the interface layers, are detected by the sensors, conducted by the geophone flute and recorded in the acquisition system for later analysis. From the calculated interpretation of these data, one can deduce several things, like the average speed of propagation, the geometry and depth of the interface, and the type of underground material. It is in fact the same method that geophysicists use to detect underground oil deposits. This seismic refraction method could be used on Mars during manned missions to detect underground pockets of water. The scientists from Paris and Brussels who co-proposed this experiment are already involved in other automatic experiments to be flown on the Franco-US mission NETLANDER in 2007.[2]

The aim of our experiment during this simulation is not to test the method (we know that it works), nor to find underground water near the Haughton crater, but to assess whether it is possible to conduct this kind of investigation in the field, in an extreme environment in EVA conditions, wearing a bulky EVA suit, with backpack, boots, and gloves. Well, believe it or not, it worked. It was exhausting but we managed to show that it is do-able.

We started with the traditional day briefing during which all the tasks of the day are distributed, while munching on energy bars. We anticipated that the lunch would be skipped. At about 10 h 30, the four persons foreseen to start the EVA (Robert Zubrin, Kathy Quinn, Frank Vizard, and myself) started to kit up. Frank observed the field operations and returned after a while. Charles Cockell, Bill Clancey, and Steve Braham, stayed in the Hab to support the EVA by monitoring the communications. Bill conducted additional observations on field operations for his research on human factors during EVAs.

We went out from the airlock at 11 h 15 to load the equipment boxes on the ATV trailer and we wandered to the same location of the dry run of two days ago. However, it took us longer to lay down the geophone flute with the bulky EVA suits.

---

[2]Unfortunately, this NETLANDER mission was cancelled in 2003 for budgetary reasons.

Unloading the ATV and installing the geophysics experiment equipment. *Credit* FMARS-2 crew

At some point, my helmet was totally fogged up, and as I could not clean it with my nose, ears, or head, I just had to suck some water from the tube and to expel it off on the inner surface of the helmet while leaning forward to wash off the condensation. Not very hygienic, but it worked. Later on, it started to rain and a westerly wind started to blow. A few other mishaps happened, e.g. when running the setting test on the acquisition system, I could not bring the screen up to a visible brightness level. Stupid and so simple to do in the lab or in the office, but try to do that with thick gloves (like ski gloves), in the rain and the cold with a helmet half fogged up. For my defence, I have to say that activating the keys on the keyboard with gloves was not possible and I had to rely on an additional tool, a small screwdriver, to press the keys. This took us quite a while, but we could eventually complete the experiment. We ran the three series of tests, with the mini-earthquake generated in the middle and at both ends of the geophone flute. Each test involved ten shots with the sledgehammer. MIT Geophysicist Kathy Quinn mastered the art of hitting the hammer and performed this chore with success, each time managing to miss the trigger geophone.

Performing the geophysics experiment: from right to left, Kathy Quinn hits the slammer hedge, Robert Zubrin supervises and Vladimir Pletser check the data acquisition unit. *Credit* FMARS-2 crew

After successfully completing this series of tests, it was time to pack up, avoiding knots and entwinement in the one hundred metres of electric cabling. We went back home, to our Martian Hab, completely wet from the rain and the sweating and cold but happy of this first success. The EVA outing eventually lasted for 4 h and the whole EVA activity, including donning and doffing the suit, took more than 5 h. It was really exhausting but extremely interesting. I cannot wait to go back for a next EVA, which probably will happen on Saturday.

A word about our life in the Hab. This morning, we woke up to a strange smell in the Hab: the toilet incinerator did not work properly overnight and the toilet overflowed. So, someone (not me, luckily) had the unpleasant task to clean up the mess and fix the plumbing. It is again a trivial story and I apologize for that, but it shows an important aspect of human exploration. Wherever human beings would go, they would carry with them their daily problems, even the most unpleasant ones, and they would have to deal with them, like a crew would have to do on board a Martian spaceship. No way that you could call the plumber between Earth and Mars: you have to repair it by yourself. We could have interrupted the simulation and left the Hab while waiting for it to be cleaned and repaired. But no, we decided to stay and to continue the simulation following the rules. Nobody leaves without EVA suits and there is no external help, you do it yourself. The DIY capabilities would certainly be very high on the list of requirements for candidates to a first manned Mars mission.

The long EVA having consumed most of the day, the rest of the time was shared between writing scientific reports, daily diaries and other reports, and interviews for the *Discovery Channel* and Frank Vizard. Our dinner was prepared by British chef Steve Braham, and he did a good job with a combination of couscous with vegetables and spaghetti with a chicken white sauce. Surprising but excellent when you are starving. The weather is awful as a storm is blowing with heavy rain and some snow is forecast for tonight and tomorrow. But, moral is high on arctic Mars. With friendly Martian greetings.

Vladimir

PS: If you did not find the reasons for the colours of the Martian flag I was telling you about yesterday, here they are: blue represents Earth, mankind cradle; red is for the Mars planet of course; and green is the colour that Mars would eventually take after Terraforming, giving a breathable atmosphere to the red planet. Another reason: this colour combination was chosen after Kim Stanley Robinson's novels 'Red Mars', 'Green Mars', and 'Blue Mars', that I recommend.

Friday 13th July 2001, day 3

Third day in the Martian Habitat. It was a quiet and relaxing day. Nothing you can do against the arctic weather: it was sleeting and raining all day long. No way to go out. We had to stay inside. We had hopes to carry out a short EVA to the weather station down by the airstrip to replace some electronic components but in view of the amount of rain falling and the increased level of the river, that idea was postponed until better weather. So, we cleared up a little bit the house. I took some photos of our Hab to better explain how we live with six persons locked in the Hab.

In the Hab, second floor, left: the living room with the kitchen corner and the sleeping rooms; right: crew working and interacting. *Credit* FMARS-2 crew

Two persons could take a shower today, Steve and Charles, being the longest in the Hab, were the lucky ones today. As one of the human factor experiments to which we participate, is to assess the overall water consumption of a confined human crew of six, we were asked to spare the water and to use only what is deemed essential. There are no restrictions for the direct consumption of water to drink and prepare food; we have to be cautious for the rest. This means that we wash with a glass of water in the morning and that is practically all. The showers for the rest of the crew would be for another day, and again, it would be with cold water (understand: melt ice or snow) and Navy-style showers, i.e. under the shower, water on, get wet, water off, soap up, water on, rinse off, water off. *Et voila*! Anyway, one gets used to this style of living very quickly and it is still bearable: the constant day light, the occasional washing up, the lack of fresh food, the confinement, and the living together. Luckily, we are getting along well and there is no interpersonal conflict among the crewmembers.

I spent the morning checking the results of the geophysics experiment we did yesterday and to send the data to my colleagues in Brussels and Paris by e-mail, still from Kathy's laptop. It looks rather interesting, although we did not find any water under the Haughton crater.

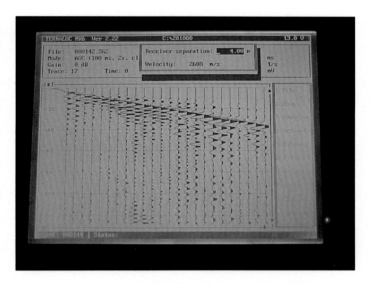

View of the screen of the Terralog, the field computer for the geophysics experiment. *Credit* VP

The afternoon was spent reading and finishing reports. Robert Zubrin insisted on playing Martian chess, a combination of normal chess, card game, and dice throwing. We did not try yet, but he slaughtered me at normal chess. I am not a master player, just an interested amateur, but I did not see anything coming during that game.

In this morning briefing, our Commander Zubrin decided that, in view of the good performance of the crew these last three days, we would be rewarded by watching a DVD tonight. The crew's choice fell on "Vertical Limit" with Sylvester Stallone. As I do not want to miss the beginning of the film and the meeting planned just before, I'll finish here this short report. So, despite the weather, moral is still going well and we are looking forward to another EVA day tomorrow.

Vladimir

Saturday 14th July 2001, day 4

Fourth day in the Martian Habitat. After yesterday's forced quiet day due to bad weather, we had a social family evening. We had first a meeting regarding some safety and health issues regarding the toilet that is still causing problems. Yes, it is not over yet. But a number of measures were decided. This incinerator toilet is not designed to be used by six persons and it appeared that the liquid production is too large to be handled by the incinerating system. So, we built a urinal linked to an emptied fuel tank and everybody uses his/her personal bottle for…, well you see what I mean. That seems to work. Then, the problem of paper bags for this incinerator. Normally we use a paper bag made of moistened paper that helps to burn solid waste. Here also we had a problem as we ran out of these special paper bags. Instead we are now using normal paper, which increases the risk of fire. But again, it works. We were not going to stop space exploration of Mars for plumbing and toilet problems. But again, it shows the importance of this facet of human life. I start to suspect that the first Roman engineers who invented the public urinals and later on the sewerage system to be benefactors of humankind. But one can still ask the question: which tortuous ways science must take to progress?

After getting a little irritated on these subjects, we could relax watching the film "Vertical Limit". The projection of this DVD on the circular wall was somehow strange as it was still daylight outside although close to midnight.

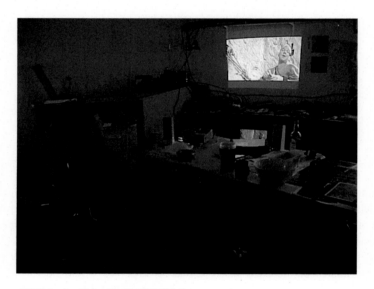

Watching a DVD in the Hab. *Credit* FMARS-2 crew

I brought with me a box of Belgian chocolates and it was an excellent occasion to demonstrate once more that Belgian chocolates are the best in the world and that they can be appreciated even in the arctic, watching a good movie. I suppose that it did help also to relax the atmosphere.

Today was again a busy day. This morning we conducted our third EVA. Although initially planned to start at 9 h 30, it was delayed by nearly 2 h due to power problems in the Hab and shortage of portable radios for the EVA. Electricity is produced by a generator working on fuel that you have to refill regularly. Steve, our engineer for the technical aspects of the simulation and of the Hab, is in charge of pampering the generator that sometimes, like this morning, has its own way. A problem of overload and fuses, eventually solved.

Our EVA was a motorized four-person EVA. Our biologist Charles Cockell was the EVA Commander, Bill Clancey, Kathy Quinn, and myself were the other crewmembers. We used four ATVs, accompanied by Joe Amaralik, the Inuit bear hunter and his shotgun, in case we would encounter some local wildlife.

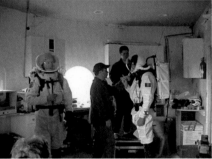

Getting ready for the EVA, we dress in turns as there are not enough room in the EVA preparation room. Left: Bill Clancey helped to don his suit; right: communication radio check in the living room. *Credit* FMARS-2 crew

Kathy Quinn and Vladimir Pletser wishing each other good luck for this EVA. *Credit* FMARS-2 crew

The goal of this EVA was to deploy some cosmic radiation dosimeters and to collect microbial samples from inside the Haughton crater. We ventured inside the Haughton crater, anticipating to be blocked by the mud at some point, but we were lucky to reach our destination, Trinity Lake near Breccia Hill in the crater, without stepping down from our ATVs. The dosimeter deployment was successful. Some were installed under about 30 cm of breccia rocks, some deployed on the surface, some near the lake, and some in the lake at a depth of 30 cm.

Charles Cockell and Vladimir Pletser installing the dosimeters in the Haughton crater under the breccia rocks. *Credit* FMARS-2 crew

It is important to be able to measure correctly the radiation exposure on Mars, as Mars does have a very weak magnetic field, practically negligible compared to the Earth one. On our planet, the natural magnetic field protects us from these radiations by deflecting energetic solar and cosmic radiations. On Mars, as the Martian magnetic field is so small, radiations fall practically without being deflected, which has important consequences for the evolution or extinction of any potential form of primitive life and for the protection of the first human crews. In addition, Devon Island being so close to the North magnetic pole (only about 200 km away), it is also interesting to measure the cosmic radiation intensity reaching the surface of the Earth close to the pole, where the lines of the magnetic field come out of the planet surface and offer slightly less protection against radiations.

The other goal was to collect a series of microbial samples living in rocks and in lakes to further analyse them back in the Hab. On our way back, we took advantage of the EVA expedition to scout the terrain for our second deployment of the geophone flute. We visited four sites, three of them in the crater: the first one close to Trinity Lake, the second one at the bottom of a small valley at the intersection of two small rivers, the third one on the inside rim of the crater, and eventually the fourth on the crater external rim. We could not decide on any of them for the moment and we are evaluating other options. We have also to consider the weather. As you know by now, it is an important factor in the Arctic. The weather today was extremely windy from the South, at 60 km/h with a protected temperature of 4 °C (so the chill factor must be added…). The weather forecast looks better for Monday than for tomorrow, and as the geophone deployment in EVA suits is quite tiring, we could decide to deploy on Monday, which would leave us tomorrow to conduct another exploratory EVA. Furthermore, we would like to make the geophysical sounding of a pingo, a mass of water ice in the ground. But we are not yet sure if there are any at a reachable distance by ATV. So, we are still debating the issue.

This evening, we are promised a shower. Yahooo! The first one since I left Resolute just a week ago, and it is about time. Also, if all reports are completed in time, we will watch another DVD this evening, so that is why I will finish this report early this evening. Our Commander Zubrin will cook dinner this evening. Apparently, a surprise, but we expect the worse!

Moral is still high (although we will see after dinner…), the remaining Belgian chocolates are excellent, and I'll sign off to run under the shower.

Vladimir
Sunday 15th July 2001, day 5

Fifth day in the Martian Habitat.

Good news! We survived the Commander's spaghetti sauce last night. We did not watch a DVD yesterday night. Instead, we had a very interesting battle between humans and the power generator. Steve Braham, our Chief Engineer, was shuttling back and forth between the Hab and the power generator trying to reset it each time it failed, which was about every 15 min. We feared that the generator would eventually win and we were about to take bets. But eventually adding some fuel did the trick. Yes, the fuel tank was simply empty. Ah! Technology!

Our fourth EVA of today was a three-person expedition which lasted two and half hours. Commander Zubrin, MIT Geophysicist Kathy Quinn, and myself went on a scouting expedition, accompanied by Joe, the Inuit bear hunter, to try to find out new potential locations to deploy the geophone flute for our geophysics experiment. We went to the *Von Braun Planitia*, not too far from the Hab, although it took us 30 min to get there with the ATVs and crossing the river. We found two potential locations, which were neither too muddy nor covered by too many loose pebbles and rocks. Pushing it a little farther, we came to the end of the *Von Braun Planitia* and continued our exploration by crossing another river and trying to climb the ridge. As Commander Zubrin let me to lead as I am supposed to be the expert to assess which place would be suitable, I tried first to climb through the snow patch and the rocks, but the ATV did not hear it the same way and tried to kick me off. So, we decided to turn back and return through a safer way.

Leaving for the motorized EVA, left: from left to right, Robert Zubrin, Kathy Quinn, and Vladimir Pletser; right: Joe Amalrik, the Inuit hunter and his shotgun posing between Robert Zubrin (left) and Kathy Quinn. *Credit* FMARS-2 crew

During the debriefing, we discussed the merits and disadvantages of all sites visited yesterday and today, and we suggested to perform the seismic experiment at one of the locations in the Haughton crater, the one between two little rivers at the bottom of a valley. Several reasons pushed us to that decision. Firstly, all the sites visited are either too muddy or covered by too many loose rocks. Secondly, the area is mainly made of dolomite (a sort of carbonate and magnesium rock), so repeating the measurements that were done a few days ago in another area would not bring any new data. Thirdly, we could not find for sure any clues to pingos or ground ice in the *Von Braun Planitia*. And fourthly, the accessibility by ATVs with a trailer carrying 130 kg of instruments is also an important factor; currently the crater is more accessible than the surrounding area. Furthermore, measuring seismic data inside the crater is also quite appealing as there would be some interest for human crews to measure underground structures in certain craters on Mars. So tomorrow, it will be on to the crater.

We would like to conduct the measurements with the two kinds of seismic sources: the already used sledge hammer and the thumper geophysical gun, allowing to shoot shells vertically down in the ground, generating the needed mini-earthquake. It will be a very long EVA expedition, certainly 5 h or more. So, this evening, early to bed. As we could not watch a DVD yesterday, we will watch it tonight. And guess what? It will be "Mars Attacks", to stay in the mood.

"Mars Attacks" is a cult movie and was so much enjoyed in FMARS. *Credit* Warner Bros

So, all in all, everything is going well. I am a little disappointed since so far, we did not spot a single polar bear. But who knows? There is another two days of EVAs and expeditions to go. So, let us hope.

Charles is cooking dinner tonight and it will be rice with chilli con carne, and for dessert, a mix of Belgian chocolates and canned fruits. Why not after all? On Mars, let's do like Martians would do, and in the Arctic, like iced chocolates… Signing off from a very foggy and humid Haughton crater.

Vladimir
Monday 16th July 2001, day 6

Sixth day in the Martian Habitat.

After the "quack, quack" of "Mars Attacks" last night, we kept on quack-quacking this morning showing that the mood was optimal for the most important EVA of our stay. It was also the most ambitious. We planned to deploy the geophone flute in two perpendicular directions in the Haughton crater, and to conduct six series of measurements, including ten shots with the sledgehammer in stacking mode and one with the geophysical gun at each of the six locations. This was to be executed by a four-person crew during an EVA of at least five hours. Kathy Quinn gave her place in the crew to Charles Cockell, as his knowledge of ways around the crater could be useful. The other three persons were Robert Zubrin, Bill Clancey, and myself.

After having cooked a warm breakfast (that is throwing the contents of a corned beef can in a pan) and swallowing everything, we started to prepare at around ten o'clock.

Getting ready for the EVA, left: Kathy Quinn helps Robert Zubrin and Charles Cockell to suit up; right: Steve Braham with Bill Clancey (right) and Vladimir Pletser both very confident. *Credit* FMARS-2 crew

I have asked by radio that Camp Manager Joe Schutt review with me the procedures to use the geophysical thumper gun, using the same shells as the shotguns. After having loaded the trailer, we were all ready to go at around 11 h 00. Charles was leading the way and I was riding the ATV with the trailer and the 130 kg of instruments in second place, followed by Robert and Bill. The weather was cloudy and drizzling. Not really ideal, but on Mars also, we would not have nice weather every day. We expected to find some spots of mud and the instructions given to Charles and I were simple: respectively avoid mud and above all, do not stop in the mud!

Yes, easy to say. Well, we did find some mud *en route*, but it is the mud that stopped us. An enormous pond of mud, invisible from far away. As I was driving this ATV as fast as possible trying to avoid getting stuck in the mud, my speed gradually and desperately decreased in this huge pool of mud. But to the defence of Charles who was leading the convoy, the area looked dry from afar. So, Charles managed to pass but not my ATV with the trailer and the 130 kg. Coming to a stop, I could feel the ATV and the trailer sinking in the mud. I came off the ATV and immediately sank as well! Up to the knees! Unbelievable! I had more and more difficulties in moving, as this mud was so sticky. I immediately thought about the sucking of quicksand. Exactly the same effect! Luckily the others on their single ATVs, after escaping a similar fate, came to my rescue.

Falling up to the knee in the arctic mud, left: Robert Zubrin came at my rescue; right: finally, up on my own feet. *Credit* Discovery Channel

After debating on what to do, whether we should pull the ATV and the trailer with the other ATVs or unload the trailer and leave the boxes in the mud, we tried every possible combination of pushing and pulling by hand and by ATVs but to no avail. It was appalling. Each one of us in turn was falling in this mud and had to help each other to extract ourselves from this viscous and sticky coating of mud. After more than an hour of falls, of miry discussions and muddy trials, we had to seek external help. John Schutt and the team of the *Discovery Channel* accompanied us, the first one with his shotgun, the others with their cameras to record for posterity this Berezina. John finally threw us two rolls of ropes that he had the intelligence to take with him and he suggested attaching two ropes to my ATV, now sunk in up to the seat, and to pull it with the three other ATVs. The trailer had sunk up to the bodywork.

The trailer with the 130 kg of equipment sank in the mud and could not be pulled out by two or four people with muddy and wet EVA suits. *Credit* Discovery Channel

Everybody was in the mud up to mid-thigh. I asked myself why I was not sinking any deeper and why my feet were so cold. And then, I understood. First our EVA suits were not waterproof and water was infiltrating inside the suit and in the boots. And second, the reason why we would not sink any deeper was simply that we were standing on the layer of permafrost, the layer of ground permanently frozen under polar latitudes. Luckily enough, the permafrost layer was not any deeper as we would have sunk even deeper.

Eventually, combining our pushing efforts and the pulling of the three other ATVs, everything started to move, slowly. My ATV came back to the surface while slowly moving forward to finally stop on firm ground, a few metres farther. Meanwhile, after having pushed at the back of the trailer, I fell on all fours in the mud and once again, it was impossible to move. As my cable antenna broke off the radio box during an earlier fall, I was off communications: to call was impossible and any way, without hands, I could not activate the radio push button. Once again, I felt myself being sucked by this mire, but this time I was on my hands and knees. Fortunately, one of my companions still standing saw me and came quickly to lend me a rescuing hand. Once out of the mud and back on solid ground, we reviewed the situation. It was not glorious. We were all covered with mud, the helmet nearly completely covered.

Pulling the ATV trailer out of the mud finally worked with additional ropes (left). Robert Zubrin nearly unrecognizable with his helmet covered with mud. *Credit* Discovery Channel

Without radios, we could no longer communicate by voice nor by signs; we had lost more than an hour and a half and we were exhausted from this battle with the arctic mud. Furthermore, the weather was getting worse and worse. Shouting to each other through our mud-covered helmets, we eventually managed to make the only possible decision: to abort the EVA expedition and to return to the Hab. I felt most unhappy and frustrated by the situation, as it would mean to renounce to conduct our experiment and to come back without data and without scientific results. I do not like that and I hate giving up. But considering the conditions and the circumstances, I had to recognize that it was the wisest decision to take. We were hoping to come back to the Hab, to get warm and to get some rest.

But it was not to happen so easily. Having turned back, following another way and driving cautiously because of the dried mud covering our helmets, the ATV with the trailer that I was driving again got stuck in another pool of mud, again invisible from afar. This was unbelievable! A real nightmare! The last time that a similar situation happened to me was in Africa twenty years ago, where we got stuck several times with motorbikes and jeeps in flash rain and laterite mud. But here in the Arctic, nobody could have imagined that we would have so much rain in the last weeks to the point where it was difficult to move around.

Once more, knowing the tune by now, we eventually managed to pull the ATV and the trailer out of the mud, again in EVA mode, that is still wearing our space suits. Again, pushing the trailer, I fell on my hands and knees in the mud, but this time I had mud up to my shoulders and hips being on all fours on the permafrost in fifty centimetres of arctic mud. Luckily once more, one of my fellow crewmembers could help me out of it, as on my own, I would not have made it out. And once more, we left by another route.

Finally, we made it home, to the Hab, exhausted, after three and a half hours of a muddy EVA. Once inside the Hab, we burst out laughing seeing each other transformed into living statues of mud. Kathy and Steve, who stayed in the Hab and worried to be without any news by radio, looked at us as if we were becoming mad. It took us another half hour to take off all the suits and the undergarments totally soaked with mud and icy cold water.

The four crew members back in the Hab from a very muddy EVA (left) and taking the suits and the mud off in the ESA preparation room (right). *Credit* FMARS-2 crew

Slowly and step by step, we got warm again, washed ourselves, ate something, and started to feel like humans again.

Andy Liebermann, the *Discovery Channel* team leader, was exulting. He filmed the best images of the week. No doubt that he would exploit this disaster and show it as the story of a crew lost during an expedition in the Martian mud (although there is no mud on Mars…). We decided to give ourselves 2 h of rest before debriefing. Meanwhile, the *Discovery Channel* guys interviewed us individually to have the story told by each of the survivors of this unforgettable expedition.

At the debriefing, still being unhappy of the outcome of this EVA, returning without any results, I remarked that we could have chosen another route, or requested outside help sooner which would have allowed us to continue. But what was done was done, and there was no point in rehashing the past.

Post-EVA debriefing, nobody smiles; from right to left: Robert Zubrin, Bill Clancey, Vladimir Pletser. *Credit* FMARS-2 crew

Nevertheless, everybody agreed that it was the most difficult EVA outing, that nobody could have imagined so much hidden mud and that it was the worst muddy experience that happened so far in all arctic campaigns. Of course, there is no mud on Mars, and as such, this simulated EVA was not representative of a Martian activity or of environmental conditions that astronauts would encounter on Mars. However, from every experience, positive or negative, there is always some conclusions to be drawn, some lessons to be learned. In this case, the way that the group managed to function in such adverse conditions and the interactions among the four EVA crewmembers could be monitored by Bill Clancey, from inside the group. No doubt that a human crew on Mars would have to face critical field situations, and the way that we acted in unexpected and repetitive situations allowed to pinpoint several important aspects: the breakdown of communications with the control centre of the Hab and among us, the need to improve the helmet visor (try to brush up a muddy visor with muddy gloves and you still see nothing), the decision chain several times broken, etc. All this would have to be analysed in depth in the coming months. So, in that sense, it was an instructive day. But unfortunately, not for our geophysics experiment. Anyway, to give it an additional chance, we decided to try again to deploy the geophone flute tomorrow, weather permitting, on Hayes Ridge in front of the Hab to complete the three-dimensional characterization of the underground structure of the crater rim. Too bad for the crater. It is still raining and the fog thickens.

With very muddy Martian greetings.

Vladimir
Tuesday 17th July 2001, day 7

Seventh and last day in the Martian Habitat.

This is my last entry in this Mars simulation diary. It is just midnight and I am back at Resolute. The sky is bright blue and the midnight sun is still shining high. Again, what a day! It seems to me that every day spent in the Arctic is exceptional, in one way or another. And what splendid weather today! As an old Inuit saying, that Aziz just invented, goes: "If you don't like the weather, wait 5 min: it will change". Well, it has certainly changed, compared to the miserable drizzle and mud we had yesterday. But let me start back from the beginning, which means yesterday evening.

After having finished our daily reports, the entire crew was feeling a little bit down and exhausted after yesterday's disastrous EVA expedition in the crater mud. The idea of repeating our geophysics experiment on Tuesday afternoon comforted me: after all, all was not lost. We treated ourselves also to spaghetti with a tin canned salmon sauce prepared by Bill Clancey and we sat down to enjoy Monty Pythons' search for the "Holy Grail". Although relatively old, this DVD made us laugh again.

This morning, waking up to a blue sky, it was time to fill in our psychological questionnaires from NASA and from Dr. J. Lapierre, a psychologist from the University of Quebec. We started full of enthusiasm but after an hour, we realized that it would take us far too long into our last day to complete them. Furthermore, we were informed by radio that there would be only one plane from Resolute to Devon Island and back in the coming days, and that it will be this evening at 6 h 00. Isn't it ironic: bad weather, the planes can't fly; good weather, the planes are suddenly all busy and they cannot come to Devon, as they have to service other arctic stations. So, as I have to catch a plane from Resolute Thursday morning at 4 h 30, I had to be on this Devon-Resolute plane.

So, suddenly it was all rush, to pack the bags and to prepare for the afternoon's last EVA. We left our questionnaires for later and we concentrated on cleaning the EVA suits from the dried mud.

The afternoon EVA was a three-person EVA with Robert Zubrin, Kathy Quinn, and myself and lasted two and half hours. We went to Haynes Ridge in front of the Hab to deploy the geophone flute in the direction perpendicular to the crater rim, to complete the three-dimensional sounding of this part of the crater rim, after last Thursday's first part. It was rather warm, up to 8 degrees Celsius (above zero!) and again we were sweating and exhausted from the heat this time. Some journalists and cameramen from *CNN* just arrived and we put on our best smiles, although it would probably not be visible on TV, and we managed to finish all the measurements in time in front of the *CNN* cameras. Data were saved and I will send it tonight by e-mail to my colleagues at the *Institut de Physique du Globe de Paris* and at the Royal Observatory of Belgium for further processing. We came back in time to pack all the instrumentation to return to France and to prepare for leaving the Hab. The three crates of geophysical equipment were still full of mud from our expedition of yesterday and I would have no time to clean them before loading them into the return plane. I hope that my colleagues in Paris would understand the difficult conditions in which we had to operate here.

The new crew has already arrived on Devon Island Monday night, and the changeover was foreseen at 21 h 00, except for me who left immediately for the base camp to catch the six o'clock plane. What a strange feeling to walk outside for the first time without the bulky EVA suit and helmet, and what a mixed feeling of joy and sadness to leave the Hab to return to civilization (to base camp first) and the crew with whom I shared this unique experience.

At the base camp, I met with Dr. Pascal Lee, the Chief Project Scientist of NASA AMES Haughton Mars Project, and we discussed potential future collaboration, which sounds rather interesting. I shared the base camp evening dinner, a giant Mexican omelette with green peppers, black beans, and salsa sauce. We exchanged views with the new crew on the Hab experience. And suddenly, it was time for me to leave Devon Island. The plane was there and all the equipment was on board. Scientists, doctors, cooks, camp managers, everybody helped to carry the loads and bags into the plane, the only link between this unique island and the rest of the planet. I shook hands with everybody and we exchanged our good byes and our wishes of staying and of leaving, and we took off into the infinite blue sky.

What a beautiful place seen from the plane, with a breath-taking view of this arctic desert with snow patches and this blend of brown and grey colours. Already the Hab was left behind, a little white dot on the rim of this huge crater, so similar to Mars in its desolate and scarred aspect and at the same time so hospitable unlike Mars.

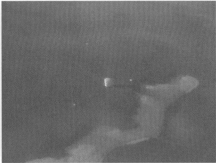

Two views of the Hab from the Twin Otter plane leaving Devon Island for resolute. *Credit* VP

Flying above the frozen sea, we spotted a huge iceberg trapped in the sea ice. The flight lasted 45 min as it did a week ago. After landing at Resolute, we were greeted by Aziz and Colleen Lenahan, the NASA HMP logistics manager in Resolute. Temperature was a summer high of 12 °C, nearly unbearable.

Iceberg in a sea of ice from the returning plane (right). The sea of ice under the sun from the plane. *Credit* VP

I was looking forward to my first hot shower at Aziz' hotel and it was even better, as there was a Jacuzzi bath in the room. Yes, I confess: while I was relaxing in the bathtub, I had a thought for my companions who stayed at the base camp spending their first night out of the Hab and for the new crew starting their simulation with the same enthusiasm that we had a week ago.

Simple pleasures in life take on another dimension when you have been deprived of them for some time: a warm shower, soap, and tonight a real bed, after nearly two weeks of sleeping bag, camping, and sponge baths in the Hab. But don't think that I have softened; I could have continued for some time this Spartan regime. But it is so good when it stops!

All in all, thinking about these two weeks, it was a great experience. I met so many interesting people and we went through so many different experiences, sometimes difficult, but always rewarding. One always comes back richer of new experiences after an adventure like this.

A long trip awaits me now for the next three days. After Devon-Resolute tonight, Thursday morning, I will start my return journey to Europe: it will be Resolute-Yellowknife Thursday morning at 4 h 30 in the morning, then Yellowknife-Edmonton. Then the whole day in Edmonton waiting for the plane Edmonton-London on Thursday night and finally the plane London-Amsterdam to arrive eventually Friday evening. Next week, I will be gone again, but on mission for ESA. I will travel to Bordeaux, France, to take part in a parabolic flight campaign organized by ESA for student experiments. Weightlessness during these parabolic flights will be a little bit like the weightlessness that astronauts would encounter returning from the planet Mars to Earth in an interplanetary journey after a stay on Mars (well, relatively speaking of course).

I am looking forward to returning one day to the arctic 24h daylight in this unbelievable place, maybe next year. Nevertheless, the final word of this diary should be a sounding "on to Mars!"

Vladimir

Chapter 3
The Arctic After—What Have We Learned from This Simulation?

Well, a lot of things. During the two months of the short Arctic summer, six crews took turns in the Hab and numerous experiments were conducted. It would be too long to summarize them all here. The interested reader can find more details in the various daily reports and scientific reports that were published every day on the web site of *The Mars Society* on this campaign. I will only talk about the lessons learned during our one-week simulation in the Martian Hab.

First of all, generally speaking, all crews have shown during their simulation that it was possible for groups of six persons of both genders, of different races, cultures and nationalities to get along and to live and work together in an extreme and demanding environment, in conditions of confinement and isolation, while rationing water and sometimes in difficult circumstances. Of course, a week is not very long, but let us not forget that it was the first campaign of this kind. The following campaigns are foreseen to last longer, increasing gradually in the duration of confinement and isolation and the related potential psychological difficulties. The second campaign of Spring 2002 foresees stays of two weeks in a new Hab. The third campaign of Summer 2002 in the Hab on Devon Island foresees a stay of four weeks. So, little by little, the psychological aspects linked to the group dynamics of a Martian mission simulation in extreme environments can be studied progressively.

Then, from an operational viewpoint, the utilisation of simulated EVA suits taught us also a lot. From an ergonomic point of view, it is obvious that most of the tools and instruments commonly in use on ground or in the lab must be adapted to the interfaces of the space suit and of the EVA gloves and boots. For example, to press a key on a computer keyboard required an additional tool like a small screwdriver or a nail. Kathy Quinn, the geologist of our rotation, adapted in a permanent way a small screwdriver by taping it to the right index of her glove, allowing her to activate and operate her GPS receiver during our EVA outings. Similarly, small mirrors mounted on the forearm of the suits allowed us to watch behind us while driving the ATVs, which helped a lot at the end of our rotation. The activation of small buttons or the installation of electrical connectors on the field computer required also another tool to be used, a pair of pliers or a jemmy in this

© Springer Nature Singapore Pte Ltd. 2018
V. Pletser, *On To Mars!*, https://doi.org/10.1007/978-981-10-7030-3_3

case. But these adaptations were quite obvious and the instruments and tools that a first Martian crew would take with them will be adapted to these kinds of interfaces where the human operator does not have any direct contact with his fingers.

The design of helmets of EVA suits must also be modified so as to allow an improved vision in the sagittal plane (the vertical plane of vision passing through the nose and perpendicular to the forehead). It was practically impossible, bending forward to see one's own pockets or the small portable radio fixed to the belt or to the breast pocket of the suit. We used a simple trick consisting of using the pockets of somebody else's suit instead of our own pockets to put used tools or collected rock samples. The design of an EVA suit helmet is most difficult as it has to be sufficiently solid and resistant, but must at the same time allow sufficient leeway to move the head and the torso as naturally as possible. But in general, the design of EVA suits has always been something complicated. Several generations of space suits have succeeded one another since the beginning of space exploration on the Russian and American sides. The main difference between the suits used during space missions in low earth orbit or on the Moon and those that will be used on Mars is the total time that the astronauts would spend inside the suits. From a few hours from time to time during EVA's in low earth orbit or on the Moon, the time that astronauts would spend in their suits would climb up to several hours regularly and practically every day during a year or more. The suit design must then certainly be adapted for a longer and more frequent utilisation. A modular concept would be preferred with interchangeable parts allowing simpler and faster repairs.

Another point is the improvement of the vision in function of climatic conditions. As we got stuck in the crater mud during this memorable expedition and mud covered our visors, some common points can be found with potential situations on Mars. Of course, the Martian astronauts would most likely not encounter water mud, but they could be stuck in dust storms. This dust could be very abrasive and damage the helmet visor, even more as Martian winds can achieve important speeds and last for quite long, weeks or months. Furthermore, condensation of carbon dioxide, the main component of the Martian atmosphere, caused by differences of temperature with the suit could also be the cause of the formation of mist on the helmet external surface creating a sort of Martian mud on the visor. A simple system allowing to clean or to brush this layer of dust/mud would be necessary, either by a simple brush mounted on the suit forearm or by a series of transparent flexible films pasted on each other and that the astronaut could peel off one by one as soon as the visibility deteriorates too much. This last system would be similar to the one used by motocross pilots on their goggles that allow them to peel off one layer after another after mud ejections.

Still regarding the operational aspects, it appeared rather soon that if we wanted to conduct exploratory EVA expeditions, we had to have not only an appropriate vehicle, but also a way to find our way back. ATVs are an excellent solution for relatively short displacements of up to a few tens of kilometres. However, to allow the first Martian astronauts to explore farther around their base, they would need another type of vehicle, like a Martian tank. The concept of the Moon rover would probably not be adequate as the exploration EVAs would last much longer on Mars than on the Moon. Ideally, it should be a self-sustaining pressurized vehicle giving also the possibility to two or three astronauts to spend days or weeks on the surface of Mars away from their base, without wearing their EVA suits. This type of vehicle would allow increasing the exploration distance to several hundreds of kilometres, which is not negligible. On the other hand, if we have already a good idea of the topology of the Martian surface, we still do not have detailed topographic maps of the planet surface. Maps at a scale of 1/100,000 are not yet currently available. However, a Martian GPS system, similar to the one available on Earth and used for navigation and positioning by ships at sea, planes, and private cars, would be an excellent method of finding your way around on Mars. Once the coordinates of the point to reach are entered in the vehicle computer, you have only to pilot the vehicle following the indications of the GPS system. For this, you need a few satellites in orbit around Mars. Therefore, it is strongly suggested to install on board all Mars orbiter spacecraft an adapted GPS transponder. This solution, simple but so efficient, would greatly ease the exploration operations on the Martian surface.

Another point about field operations. To conduct our geophysics experiment, we had very precise procedures that we had to follow step by step to choose the best locations and orientations, to install the sensors, to unroll the cables and lines, to connect the sensors, to configure the acquisition computer and eventually to conduct the experiment itself. These procedures were written in a small paper booklet of ten pages, that we had to consult all the time to respect the prescribed chronological order of operations. To hold this booklet, to read the instructions and to turn the pages in the wind and rain were not easy tasks. We have tried several alternative methods (to have a read back of the various steps on the radio by someone in the Hab, keep the pages under a plastic envelope, etc.), but none were found practically convincing. It is suggested to envision to mount on the left arm (for right-handed people) of the EVA suit a small pocket computer, like an organizer, displaying the procedure text, or any information for that matter, on a liquid crystal screen, activated by keys on a small keyboard adapted to the finger size of EVA gloves. This system is already envisioned for EVA suits for the astronauts of the International Space Station, and would be most useful in climatic conditions like those that we have encountered.

Another important point that was put forward is the problem posed by the different levels of communication and the systems of replacement in case of temporary unavailability of a first level. Initially, we should have communicated with a mission control centre based in Denver, Colorado, and all communications should have been conducted with a delay of 20–30 min like on a real mission to Mars. Unfortunately, the satellite link that should have allowed us to communicate between Devon Island and the Denver control centre was unavailable due to a jammed antenna system. Instead we used e-mail communications with the Denver control centre and radio communications with the base camp located a few hundred metres from the Hab. It was enough to ensure the control at a routine level and to remain informed of problems specific to our stay on Devon Island. On the other hand, the radio communications among the crew members and with the Hab during the EVA expeditions suffered from several technical problems mainly due to either batteries insufficiently or wrongly charged due to the ambient cold and humidity, or by bad connections or pulling out between the receiving unit and the microphone of the headset, or by setting buttons being stuck or impossible to activate with EVA gloves. While waiting to fix these problems, we had to use in the field degraded communication modes that is not optimal but sufficient to pass a minimum level of information. This approach is certainly very instructive as it is obvious that astronauts on Mars would have to face similar small technical setbacks without having to decide to cancel an EVA outing. Even without thinking about it, we naturally came to test two modes of degraded communications that seemed obvious. When close enough to each other, to bring the helmets in contact with each other and to shout to one another through helmets was sometime sufficient. This minimum communication mode could not be impossible on Mars as there is an atmosphere, even at very low pressure. The second mode was to communicate by hand or arm signals when the EVA crewmembers were far from each other. This simplified communication mode is currently used by human operators in other extreme conditions, like scuba divers in deep water, or parachutists flying in formation. We applied this hand signal mode while driving the ATVs to indicate to stop, or to turn, or to slow down. Furthermore, for our active seismic experiment, a raised arm indicated that the acquisition system was armed and that the hammer operator, sometimes a hundred metres away, should stand ready and that all other persons in the immediate surrounding should stop moving. The raising of the other arm indicated that the acquisition system was ready and that the operator could hit the hammer on the metallic plate at his or her discretion. Although very simple, this mode was rapidly preferred to the verbal warning by radio in view of its simplicity, its briefness and its immediate understanding by all Martionauts present on the field.

Still on that subject, it would be interesting also to envision to communicate not only by voice messages by radio but also by electronic messages by radio, like sending SMS instead of talking on the portable phone. This system could be coupled to the small portable computer adapted to the sleeve of the EVA suit mentioned above. For those who saw the film *Abyss*, the hero used a similar system during a very deep dive.

Regarding the experiments conducted during our rotation, well, as already said, we did not find any underground water pockets, but that was not the goal of the experiment. We could prove that it was possible for a team of three persons in EVA suits and properly trained, to conduct a series of operations to install and run an active seismic experiment involving the set-up of 24 seismic sensors and hundreds of metres of cables, with more than fifty electrical connections to be plugged in, to configure a field computer and to conduct several tens of seismic measurements with a sledge hammer. These operations requiring precision and a given order of installation were very exhausting, mainly because of the additional weight (about 20 kg) and the cumbersomeness of the EVA suit.

This yields an important consequence on the scenario of a Mars mission with respect to the mode of propulsion of the interplanetary Earth-Mars travel. Two modes of propulsion are possible: either constantly propelled, or non-propelled or ballistic. The propelled mode is of course faster but one needs to bring along the necessary quantity of fuel (hydrogen or hydrazine) and of oxidizer (oxygen, nitrogen tetraoxide, or other). I will not discuss other, more exotic propulsion modes like solar sail or ionic propulsion, which would of course be feasible but not envisioned presently because of a lack of practical experience for manned missions. This additional mass of fuel and combustive would have to launched from Earth and would increase tremendously the budget of total mass to be sent first in low earth orbit, and further on in an interplanetary insertion orbit.

The other non-propelled or ballistic mode has the advantage of not needing to transport such a considerable additional mass of fuel and combustive, only what is necessary to inject the interplanetary craft into a transfer orbit to Mars, or in other words, to give the craft the initial impulsion, the remaining of the trip being done following the laws of Celestial Mechanics, by simply letting it go after the initial push. During the non-propelled part of the journey, the spacecraft and all its contents are in a state of free-fall and weightlessness prevails inside the craft: all the contents and the astronauts float freely if not fixed inside the craft, as they experience no weight. We know now that a prolonged state of weightlessness has debilitating effects on the human organism. The most important effects are a loss of minerals from the bone system and a loss of muscular tonus and volume. Bone demineralisation means that calcium and to a lesser extent, phosphorus and other constituents are eliminated naturally by the organism. The calcium loss is on average about a tenth of a gram per day, which is a lot when you know that the adult human body contains on average a total of 1 kg of calcium. The muscles also, as they do not need to fight against gravity to ensure a standing position, will shrink and loose volume and tonus. An astronaut crew after a six-month exposure to weightlessness runs the risk of losing so much bone minerals that they could not land back on Earth or on Mars without risking multiple fractures. These conditions have been studied and are still studied on astronauts and cosmonauts for many years during space missions on board Spacelab, the now defunct Mir station, and the International Space Station. Scientists investigate also what is called in space jargon "countermeasures', that is a set of measures to fight these bone mineral losses by an appropriate diet with supplemental calcium and, mainly, with a series of physical exercises. Astronauts are asked to do one or two hours of sport per day in orbit, either to run on a treadmill, or to work with extensors, or pedal on an ergometric bicycle (which is a bit paradoxical as they are already moving at a speed in the order of 28,000 km/h in orbit around the Earth!). To come back to Mars missions, a crew just landed on Mars after several months spent in weightlessness would not be able to conduct physically demanding field scientific and exploration tasks, like this seismic experiment that we performed during EVA outings, although this is one of the goals of Mars exploration missions. Therefore, there are two possible solutions if one wants to keep the non-propelled mode in view of the costs associated to the transport of fuel and combustive: either artificial gravity is somehow provided to the craft, or efficient countermeasures must be developed to allow the crew to arrive in a sufficient health state. Artificial gravity can be generated by putting the craft in rotation, hence creating a centrifugal force replacing the gravity force, either around a craft axis, or even better, by separating the craft in two distinct parts linked by a tether and putting the whole in rotation around a point more or less central to the tether. Regarding countermeasures, they are not hundred percent efficient to be entrusted with the physical health of an interplanetary mission crew. Our understanding of phenomena caused by weightlessness on the human organism improves after every space mission, but all the details of these mechanisms are not yet fully grasped and, *a fortiori*, the means to fight them are not yet fully developed. These two solutions however should probably be applied together in the case of a non-propelled mode journey to Mars.

This is already an important conclusion regarding the scenario of a Mars mission, obtained from a simple experiment conducted during a simulation on ground.

Another conclusion of this experiment concerns the composition of the crew. As already emphasized above, the field operations during EVA expeditions took much longer than in a normal environment. For our geophysics experiment, more than eighty percent of the time spent in the field was to choose the site of the experiment, the setting up of the material and the configuration of instruments needed to conduct the experiment, and only twenty percent for running the experiment itself. This means that the choice of the site is crucial to optimise the field operations and to not loose time. This was only possible because the scientific expertise was present in the field. This type of experiment cannot be conducted by untrained operators without the guidance by a specialist of the scientific field at hand. It is therefore absolutely crucial to have experts in the scientific fields that would be explored on Mars present among the crew of Martian expeditions, and not only trained operators. The scientific expertise must be present on Mars. To every man his job, and one must leave to scientists the task of conducting the scientific research in space exploration.

From a scientific point of view, the results of our experiment confirmed that the underground layers on the Haughton crater rim are made of dolomite or Magnesian limestone, a kind of rock made of calcium carbonate (a molecule containing carbon, oxygen, and calcium) and magnesium. These data were studied by students for their physics mastership thesis. This experiment and preliminary results were presented at international conferences in Houston, Stanford, Paris, and Brussels, and during some technical seminars at ESTEC. References are given at the end of the book.

We were also rationed for water, not for direct consumption but for personal toilet and hygiene. The hygiene itself did not suffer too much, but it made us more conscious of the wasting in everyday life whenever we wash our hands or brush our teeth or take a shower or a bath. Instead of letting the tap water flow freely, it would be so simple to turn the tap off and to utilize only the amount of water strictly necessary for these small needs of personal hygiene. So, by paying attention to little things like that and not restricting water for direct consumption, we could show that our crew used less than half of the amount of water estimated by NASA for a crew *en route* to Mars and living autonomously on the surface of Mars. The NASA estimation was 36 l of water per day and per person. Our consumption was on average of 15 l of water per day and per person, which is still a lot when one thinks about it, but this includes the water used for cooking, dish washing, and the occasional showers. It proves that it is possible to reduce at least by half the total amount of water to bring along for a journey to Mars. Once again, to measure a real consumption instead of taking for granted an unproven estimation is an important aspect of this kind of simulation in an extreme environment representing what astronauts would face on Mars.

Finally, some suggestions were made for the inside arrangement of the Hab. Everybody commented on the lack of "utilisation comfort" of the small rooms that were allocated to each of us. Not that we were expecting something spacious and of a comfort worth a four-star hotel, no, far from it. The size of these little rooms was not the problem, but more the inside layout and the ergonomics of the rooms were not adapted to long duration stays. Among others, cupboards, ranging shelves, and small working desks were cruelly lacking. But all this was perfectly bearable for a short period of a week. The lack of fresh food was also mentioned, but no deficiency in vitamins or oligo-elements was noted. First, the duration of the simulation was too short to effectively notice this, and furthermore supplemental vitamins were made available to everyone. Neither did we observe a degradation of the taste and smell senses, the simulation being again too short. It is well known however that astronauts and cosmonauts on board Mir and the International Space Station report a decrease of these two senses, all smells and tastes getting mixed up and becoming indiscernible. Talking about smell, everybody agreed to advise against an incinerator toilet for an interplanetary trip. Besides the obvious increased risk of fire, the frequent breakdowns, and the fact that a crew of six cannot live without such an important system, impose that another more reliable and more efficient system must be imperatively envisioned.

These are a few examples of the lessons learned during this short isolation simulation. There are many others of course and all this is under study following the six rotations that took turns in this Martian base in the Arctic.

I did not talk about psychological aspects, first as I am not a psychologist, it is not my field and I would not want to say anything incorrect. These kinds of observations, either direct (by a real-time monitoring of communications and interactions between crew members) or indirect (by questionnaires and interviews on impressions and behaviours), is a work in progress and it would be useless to try to draw partial conclusions. All that I can tell you is that we did not have any interpersonal problems and that, to the contrary, all the members of our crew had the impression that we formed a coherent group and interacted constructively, even during the two or three difficult moments that we went through.

This, I think, is the best conclusion that we can draw from this simulation. The exploration of Mars and its preparation is a human venture and as such, it reinforces the relations between human beings, men and women, sharing the same passion for this ultimate goal of discovering and opening a new world to humankind.

Part II
The Desert

Chapter 4
The Desert Before

One fine morning of January 2002, I received an e-mail from Robert Zubrin. In his characteristic direct and concise style, in less than ten words, he asked me to go again with them, this time in the desert, in the new Hab installed in the middle of the desert of Utah. I replied directly "On to Mars!" (after at least having checked the dates). I was again enthusiastic. Like before, six rotations were foreseen with mixed

View of the Mars Desert Research Station (MDRS) of *The Mars Society*

© Springer Nature Singapore Pte Ltd. 2018
V. Pletser, *On To Mars!*, https://doi.org/10.1007/978-981-10-7030-3_4

crew of six persons for stays of two weeks each. I would take part in the fifth rotation in April 2002, with Bill Clancey acting as Commander, who was already in the previous expedition in the Arctic. The four other crewmembers would be Andrea Fori, a geologist from California; Nancy Wood, a biologist from Chicago; Jan Osburg, an aerospace engineer from the Stuttgart University in Germany; and David Real, a journalist with the Dallas News of Texas. This new Hab was the second built by *The Mars Society* and deployed this time in another extreme environment, a desert! After the Arctic cold, the desert heat! This second Hab had new scientific possibilities, as an experimental greenhouse was annexed to it and some Martian biology and botanical experiments were foreseen.

Robert Zubrin asked me also if I could come back with the equipment of our geophysical experiment to repeat seismic measurements like on Devon Island. I verified with Philippe Lognonné if the equipment could be available again on loan. Unfortunately, it was already in use for training sessions of students in geophysics. As time was relatively short, I had to find another experiment. I talked about it again with Véronique Dehant who put me in contact with another French geophysicist who was taking part in the scientific team of the NetLander mission working on an experiment to measure the planetary magnetism of Mars. He was interested but, unfortunately again, the scientific instrumentation was not available for the foreseen dates. In the meantime, I had discussed with one of my colleagues at ESTEC, Dr. Christophe Lasseur, a specialist in life-support systems for space missions. He proposed an excellent experiment idea, namely to grow vegetables in the Hab and in the attached greenhouse. The idea was not really to explore the conditions and the methods of growing plants, but more to study the reactions of a crew facing the chores of gardening and to measure the psychological impact of taking care of green plants and vegetables. We discussed it at length and it appeared that it is an important question. The first crews on Mars would not be able to take with them all the food that they would consume during the two years or two and half years of stay on Mars. They will have to produce part of what they are going to

Should a greenhouse be pressurized and attached to the Hab to allow the astronauts to stroll like in a garden (left, *credit* NASA), or should it be detached and only accessible by EVAs? (right, *credit* C. Lasseur)

consume. Several questions are asked and not yet answered regarding the design of a Martian greenhouse. Should it be pressurized or not? And if so, at what pressure? At a terrestrial atmospheric pressure (one bar) or a Martian pressure (about 7 millibars)? And made of which atmosphere? Terrestrial (Oxygen and Nitrogen) or Martian (Carbon dioxide)? Should the greenhouse be attached to the Hab to allow Martionauts to access it without EVA suits or should it be detached and accessible only under EVA conditions? From a psychological point of view also, a greenhouse attached to the Hab or inside the Hab would allow Martionauts to retire to this inside garden and maybe enjoy some moments of solitude and relaxation while taking care of the plants, whereas a detached greenhouse accessible only during EVA outings would require to don an EVA suit and to perform an EVA expedition, which is a heavier procedure, longer and less relaxing.

There was material here for an interesting experiment. Not that we wanted to answer all these questions, but we might find some first indications that could orient and eventually quantify the choice of a design over another. As you can see, technical and scientific experts are working already to iron out these kinds of details that would condition the future Martian expeditions in twenty years or so.

The experiment itself would be very simple: to grow some vegetables chosen for their properties of fast germination and maturation; eventually to consume them at the end of the rotation and to ask crew members to note their comments, remarks, etc. An important point is to not let the other crew members know about the final goal of the experiment in order to obtain frank and naive answers. I nevertheless discussed it by e-mail with our new crew Commander Bill Clancey, who understood very well the goal of this experiment.

One of the characteristics of this second campaign of simulation was for the previous crews to explore this part of the desert surrounding the Hab using the GPS system. This Global Positioning System was already used during our first simulation campaign in the Arctic, but it was so useful that during this second campaign, it became an indispensable tool for the exploration of the desert. All the information gathered by the previous crews were noted and commented in a database and in expedition daily reports of previous crews, all accessible on *The Mars Society* web site. We had long discussions by e-mail with the other members of our future rotation before we actually met to prepare the approach that we were going to follow to exploit these data. Bill asked us all to read the science and the EVA expedition reports of previous crews to avoid arriving without knowing what has been already done. In other words, it was no longer a question of playing to be the first crew landing on Mars and starting to explore a *Terra Incognita* (or better an *Ares Incognita*), but more to act as a following crew arriving after several other expeditions and pushing the exploration further from the point where others before us left it. We had therefore to study all that was done by previous crews. I spent several evenings (and nights) reading and studying previous rotation reports, while keeping on working during the day on the several projects that I am responsible for at ESTEC. It was just as well we did it, as it was rather interesting and instructive. Although we hadn't been there yet, we became familiarized with the place.

Our rotation will be supported by a mission control center located in California. Our contacts with this control centre will be only through e-mails and the people who would help us electronically from far away are themselves passionate of Mars exploration, all members of the Californian chapter of *The Mars Society*.

After a while, we felt ready to face this new and exciting environment. The other members of the team seemed to me all motivated and sympathetic, which was auguring well when thinking that we were going to spend two weeks confined in a few cubic metres.

Bill asked Andrea and Jan to draw and manufacture a crew patch representing our team and our rotation, which was well done.

Everything was getting ready and I felt full of energy to attack this new phase of preparation of Mars exploration. On to Mars and new adventures.

The MDRS-5 crew patch with the names of the six crew members. *Credit* MDRS-5 crew

Chapter 5
The Desert During—Mars Desert Research Station Diary

Note

This is the daily diary that I kept during the journey to and my stay in the Desert of Utah. Large excerpts have been published on the web site of ESA, and of the French Chapter of *The Mars Society*, and daily in the Belgian newspaper '*La Dernière Heure*'. This is the text in its integrality. Dates are indicated and days are counted since the day of departure.

Sunday 7 April 2002, Day 1

Martian greetings, Earthlings!

My name is Vladimir Pletser and I have the chance to take part in an extraordinary adventure as a member of the fifth crew of this second international campaign of simulation of a Martian mission. I hope to keep you informed about our scientific work and our daily progress in this station.

This is our first day at the Mars Desert Research Station (MDRS in short). It is a fantastic place in an unbelievable surrounding, installed in the Desert of Utah, several hundred kilometres south of Salt Lake City. It is the second of four stations that *The Mars Society* intends to deploy around the world. The first one was set up two years ago, in the uninhabited Island of Devon in the Canadian Polar Circle and it was used for the first international simulation campaign to which I participated. The third station will be European and deployed in Iceland possibly in 2004–2005.[1] A fourth station will be installed in a few years in the Australian outback. These remote places on Earth resemble by some aspects what can be expected to be found

---

[1]The EuroMars Habitat was planned to be deployed in Iceland, but, unfortunately, it was never deployed for budgetary reasons.

© Springer Nature Singapore Pte Ltd. 2018   79
V. Pletser, *On To Mars!*, https://doi.org/10.1007/978-981-10-7030-3_5

on the planet Mars, either by climatic or geologic or possibly biologic conditions, sufficiently close to conduct scientific experiments similar to what a Martian crew would conduct. The idea is to use these earth extreme environments, rightly named "Mars analogues on Earth", to demonstrate that a manned mission on Mars is feasible and that a human crew can live autonomously inside a station, familiarly called 'Hab', specially designed to resemble a future first inhabited base on the red planet.

But let us start from the beginning. Arriving here is much easier than last year in the Arctic. It only took 24 h, door to door. As I live and work in The Netherlands, I left Amsterdam Saturday morning and arrived at Atlanta nine hours later. I waited for two hours for the connection to Salt Lake City and at the airport, I was picked up "at random" by security for a strip search. Even my shoes were examined! Yes, the security checks were reinforced after the terrorist attacks of September 2001. After another four hours, I landed at Salt Lake City where I met the rest of the crew. Bill Clancey, a computer specialist at NASA, with whom I spent one week at the Mars Station in Devon last year and who will be our Commander for these two weeks, and Andrea Fori, a planetary geologist from California, were at the airport gate to greet me. The rest of the crew was doing the last shopping for the next two weeks. Nancy Wood, a biologist from Chicago, David Real, a journalist from Dallas, and Jan Osburg, an aerospace engineer from Stuttgart, Germany, filled up three supermarket carts with cans, vegetables, fruits, meat, …, enough to sustain a siege for more than a month in our Mars base. We then drove south a few hundred kilometres with two vans for another four hours, and after stopping for a meal, we arrived at around one in the morning at Hanksville, a small village, so small that it was forgotten apparently during the last American census (no joke!). If you look at a map, you will see in the southern part of Utah, three roads that form a Y; Hanksville is located at the intersection of these roads. Our base, or Martian Hab, is located another twenty kilometres or so from this village. This area was apparently the place of Butch Cassidy's feats in those Far West days of the 19th century. We stayed in a motel for a short night as this part of the US passed from Mountain Standard Time to Mountain Daylight Time, equivalent to what we did two weeks ago in Western Europe, which meant another hour less to sleep. Well, again, we are not here to sleep and time will not be an issue during our two-week simulation.

This morning, we drove the last kilometres through the most grandiose landscapes of the American West to reach our final destination for the next two weeks. The surroundings are so unearthly that the Hollywood film director James Cameron came here to see whether he could shoot some scenes of his Science-Fiction movies.

View of the Hab in the surrounding desert. Notice the stratification of different colours all around, corresponding the Cretaceous-Tertiary transition. *Credit* MDRS-5 crew

The Hab with the Greenhouse, three ATVs and the Martian flag. *Credit* VP

We met the departing crew who was partly sad to leave and partly happy to go back to civilization. I guess that we would feel the same in two weeks, but for the moment we have so much ahead of us that our minds are filled with other expectations.

The MDRS-5 crew, from right to left, Bill Clancey, David Real, Andrea Fori, Nancy Wood, Jan Osburg and Vladimir Pletser. *Credit* MDRS-5 crew

After we were briefed on the various little things to do and not to do by the departing crew, we had several engineering chores to perform. Luckily, Frank Schubert, one of the officers at *The Mars Society* who is also the general manager of this place, was here with some of his engineering team to organise everything. The greenhouse door was repaired as it had been blown away by strong winds in the previous days. A new generator was installed and functions now without interruptions. The outlet and leach field of the biolet, the biological toilet that replaces the infamous incinerator toilet of the previous simulation, still needed to be sorted out. Well, yes, it is important, as we want to run this simulation in a closed environment, to take care of these things before we actually start the simulation.

The Hab is quite similar to the one installed in the Arctic. Having the same dimensions, the inside lay-out is quite similar, with the differences that the stairs to access the first floor are placed in the other direction, that there is a skylight on top of the Hab and that there are more little electronic gadgets on the tables around the circular wall of the living room. And sand everywhere. When the wind blows, this desert sand gets everywhere.

We had a long brain storming session on the modalities of this simulation and on how the various domestic chores will be shared between the six crewmembers. It goes from cooking and cleaning the dishes to verifying the alignment of our satellite antenna and filling the water reserve. I will tell you in more details about theses chores during these two weeks.

Nancy Wood demonstrates the art of abseiling to evacuate the Hab via the roof top in case of evacuation. Jan Osburg is appreciative. David Real and Bill Clancey set up the 24 h recording video camera to log all crew movements in the Hab

I am very much impressed by the enthusiasm and the willingness of the members of this crew, and by the organization that Bill, our Commander, wants to bring in this simulation campaign for the meticulous recording and archiving of samples and data. It is true that in the three months that this station is used, an enormous amount of data on the local biology and geology and on geophysics experiments was collected and rock and biological samples accumulate in the ground floor lab. So, we need a good archiving and labelling system to not mix the things up.

We had also a few visitors who heard about the simulation and wanted to have a look. Obligingly, we showed them around the Hab and explained the goals of our experiments. We told them that they were lucky as we would start the full simulation mode by midnight tonight and that the next opportunity for external visitors would be in two weeks only.

The full simulation mode means that we will not be able to go out of the Hab without wearing EVA suits to simulate that we are on the surface of another planet. These EVA suits are similar to those that we were wearing last year in the Arctic.

We had also the chance to ride the All-Terrain Vehicles (ATV in short) that we will use during our expeditions. These ATVs are quads, a kind of four-wheel motorcycles, the same as those we also used in the Arctic. This ride at sunset was extraordinary as it provided spectacular views of the different shades of reds, yellows and greens of the surrounding canyons and hills. I do not have enough words to describe this feeling: just out of this world!

Views of the Utah Desert around the Hab. *Credit* MDRS-5 crew

All in all, these were two excellent days as I could meet the rest of the crew with whom I would spend the next two weeks in this Hab, and we had a very nice impression on how the simulation would go from the departing crew, as they did not want to leave.

The jet lag starts to take its toll rather heavily, it is 23 h 30 (and I do not know anymore at what time my body clock is). So, I will sign off here for today, with a huge Martian smile on my face.

On to Mars!

Vladimir
Monday 8 April 2002, Day 2

Martian greetings, Earthlings!

Our second day in the Mars Desert Research Station went beautifully well. We had a long briefing session this morning to discuss our different EVA expeditions that we planned for this week.

Morning briefing from two different angles: left: from left to right, Nancy Wood, Bill Clancey, Jan Osburg, Andrea Fori, David Real; right: Vladimir Pletser, Nancy Wood, Bill Clancey. *Credit* MDRS-5 crew

We decided also to change our schedule, something that a crew on Mars will always be able to do. We adopted what my colleague Nancy called the Spanish schedule that is delaying all our external EVA activities till after four o'clock to avoid the burning sun of the afternoon. Actually, the temperature differences in the desert are quite marked: the highest temperature was +32 °C (about +90 °F) at 12 h 38 pm yesterday and +4 °C (about +39 °F) at 5 h 20 am this morning.

The first EVA of this rotation was to set up the station greenhouse, located about ten metres from the Hab but still necessitating to wear our EVA suits. One of the experiments that I am taking care of with the biologist Nancy Wood consists in observing the growth of some vegetables from planting to harvesting. We have four kinds of seeds: radishes, alfalfa sprouts, arugula salad and Chinese tatsoi cabbage. Nancy and I started to prepare and rehearse the procedure that we would use with the EVA suits in the greenhouse. As part of this training, we planted the seeds in growth boxes that would be left inside the Hab to be used as comparison with those installed in the greenhouse. Nancy was quite resourceful in adapting a small lab funnel and some Eppendorf tubes (which are small lab tubes with small funnel ends) to assemble a single-seed distributor.

Left: Nancy demonstrates how to deliver seeds one by one using an Eppendorf tube and a spoon. Right: Nancy and Vladimir prepare the growth boxes to be left in the Hab. *Credit* MDRS-5 crew

So, the plan was ready and rehearsed. We started to kit up, following the EVA suit donning procedure, and as it was the first time for four of my fellow crew members to don these simulated space suits, it took us more than an hour to go through the details of how to avoid your hair getting caught, or your nose being punched by the helmet, or your earphone falling of your ear once outside.

Left: preparing the EVA suits and helmets, we were soaping the inside part to avoid fogging. Right: Nancy and Vladimir shaking hands before the EVA. *Credit* MDRS-5 crew

So, ready we were and off we went, having checked that we had all the plants, the growth boxes and the seeds, but obviously forgetting part of the tools to be used. After having waited five minutes in the airlock to simulate the decompression, we reached the greenhouse. Once inside, we had to improvise with what we had to complete the plan. Obviously, we had to deviate from the rehearsed procedure that we were so proud of a few minutes ago. This shows clearly that nothing can replace a field trial in the most realistic conditions possible, and that despite all imagination that one can use, there is always a detail that escapes the eye of the scientist or the engineer. This kind of simulation is therefore necessary already today if we want to reach and work on Mars, in a few years, say twenty years or so.

Let me tell you that it was not easy to plant seeds of about one millimetre diameter, one by one, using simulated EVA gloves (less flexible than ski gloves) in a greenhouse overheated by the sun. Nevertheless, in a little more than an hour, we were done: all the seeds were planted and watered. Job done!

In the Greenhouse during the first EVA, left: Nancy succeeds in opening plastic zip-lock bags containing seeds and tools, with EVA gloves; right: Nancy manages to deliver seed one by one using the funnel tube in the rock wool mat. *Credit* MDRS-5 crew

Upon return in the Hab, we debriefed our colleagues and we let them go on the second EVA of the day, which was a pedestrian reconnaissance EVA for places exposed to the wind. More on this tomorrow.

A word about our everyday life on board. We also have to manage this base as a house inhabited by six people, which means that all domestic chores, technical maintenance, etc. need to be done. We decided to rotate these tasks every day or so, alternatively each crewmember would have to do something different. I volunteered to be today's DGO, i.e. Director of Galley Operation, that is the poor soul who does all the kitchen chores: cooking, cleaning, dish washing, garbage emptying, etc.

So, I prepared a Martian meal for my fellow Martians-to-be, and it was not easy. Try to prepare a four-course meal for six people with only two heating plates and the entire power system going down as soon as you try to use a third appliance. Well, believe it or not, I manage to prepare some *Valles Marineris* delicacies (i.e. salmon toasts), an alien soup (he… a mixture!), Martian rice with pork chops *façon Olympus Mons*, and a *Lycus Sulci* fruit salad. We have found in a kitchen cupboard a little bottle of an ultra-hot and spicy sauce from Costa-Rica, left over from a previous crew (and aptly called "Ass in space"). We have tried it with the Martian rice. Well, no comment on this, except that it was an interesting experience….

So, voila! A busy day that ends well. I have finished my DGO duties at around 10 h this evening, in time to write my daily reports. The Commander just called us to watch a summary video of last year's FMARS activities on Devon Island. Good memories and some good laughs also, but still useful to understand and analyse the differences in the environments and how it will affect our work.

The plan for tomorrow is to do a scouting EVA with the ATVs as long as the weather is mild (it was cloudy this afternoon and evening).

So, I will sign off here, before the traditional power cut to refill the generator every six to eight hours.

So again, with a huge Martian smile, I wish you all a Martian good night and I sign off for this second day.

On to Mars!

Vladimir
Excerpt from the daily activity logbook

Our Commander Bill asked us to hold a logbook of all our daily little tasks and actions indicating the time when we started and finished them. This is one of the experiments that he is conducting and that would allow him to build a better model of activities of a crew on Mars, in order to understand where time goes when you have so much to do and how it can be improved. So here is an excerpt for the first Monday. It is not very interesting to read but it is nevertheless instructive as it gives you an idea of a typical day schedule.

Yesterday, I was the DGO and I worked quite a lot. Woke-up at about 7 h 30, breakfast from 8 h till about 8 h 30, then as DGO, I cleared the table and the kitchen. Briefing from 9 h to 10 h (we decided to limit the morning briefings to one-hour maximum: whatever was not said today would be said tomorrow, if ever) and then, we started our experimental work, each on his or her side.

At around noon, I left Nancy alone to continue the preparation work in the ground floor lab while I started to prepare the lunch, that is put everything on the table for a few sandwiches quickly swallowed with coffee for some, and tea for the others. Again, I cleared up everything and at around 2 p.m., I went back to the lab to prepare for our EVA outing. The afternoon was spent working, which included the preparation of the EVA, the EVA itself, the return from EVA, the debriefing, the notes to be taken, and then put everything back in place and start the daily reports.

At 18 h 15, I went up to start the dinner and do the washing of the day. I prepared a four-course meal with only two electrical plates. When I tried to use the toaster for the entrees, the power generator went off. So, all in all, it took time. At 19 h 45, we sat for dinner and around 21 h, I cleaned up and washed the dishes until 21 h 40. After having taken the garbage down in the back airlock (they would be disposed in an appropriate place during a next EVA), I came back in my little room to write my diary. Bill called us around 22 h 25 to watch a video until around 23 h. Then, I finished my report in English, translated it into French and I finished at around half past midnight. I searched through all the digital photos taken during the day to choose four to accompany the text to be sent. I transferred everything on a diskette and I sent it via the Internet around 1 h 15. I brushed my teeth and crashed down into bed at around 1 h 30. It was a long day, the day of a DGO! I woke up twice to go to the bathroom, around 3 h 30 and 7 h 20. The sun was already up and I went back to bed to read until 8 h. Then a quick sponge-shower and breakfast to start a new day.

Tuesday 9 April 2002, Day 3

Martian greetings, Earthlings!

Another great day in the Martian desert of Utah, USA. This third day was again very busy. But first, I want to share a special moment with you. Three of my colleagues (Bill, Nancy and Andrea) went on EVA.

Bill has his microphone connection checked while Nancy and Andrea wait in the background. *Credit* MDRS-5 crew

I am in the Hab with Jan and David. We are working on our computers, listening to an old song from the seventies from the band *America* ("A Horse with No Name") and in the background, we can hear the conversation with the hissing radio sound from our three EVA companions. The sun is shining and it is warm in the Hab. And suddenly, I felt transported by the song talking about the desert and the background EVA radio noises while I was writing to "Earth". It made me feel like I am already on Mars doing the same kind of exploratory work. Nice feeling!

Well, as said, again a busy day. An early start allowed me to finish Zubrin's book "First Landing" (excellent!) and to start another book on evolutionary biology. After breakfast, took place the traditional morning briefing where we discussed the day's activities. I would not go on EVA today. Instead my priorities would be to sort out, with the help of David, the Internet connections, which are still not functioning properly. It would be too long to explain the different systems that we are depending on, but not everything is running as smoothly as it should. David is the DGO (Director of Galley Operations) of the day and we were treated to a Martian Lander sandwich (tuna fish with red peppers, celeries and pickles) at lunchtime. Not bad. It would satisfy the hunger of any Martian explorer.

With Jan, we took care of other chores, like setting up our GPS's. For those interested, the coordinates of the MDRS Hab upper floor are UTM12 S 0518236, 4250730, elevation 1378 m, with a theoretical accuracy of 5 m (based on six satellite measurements). So, with this, we should not get lost when we are going on EVA.

Jan (standing on second floor) and Vladimir (lying on top of sleeping rooms) set their respective GPS. Believe it or not but the difference was still about several meters between the two settings! *Credit* MDRS-5 crew

Actually, all the preceding MDRS crews have relied also on the GPS system and more than 100 Waypoints have been logged at interesting places in the desert. One objective for our crew would be to revisit some of these locations to assess the possibility of finding them again and to collect other samples. It raises as well the question on how the first Martian astronauts would find their way around. They would definitely need a system similar to the GPS on Mars and it would be an interesting idea to already consider now placing GPS-like transmitters on Martian orbiter satellites that are foreseen for the coming years in order to prepare for the next step, the human exploration.

Another chore that we took care of with Jan is to refill the generator, three times per day. This obviously is an off-sim activity for safety reasons because of potential fuel spillage and unencumbered handling of the generator. Explorers on Mars would not have to do this: they would rely either on small nuclear reactors or on generators fueled by methane produced from the Mars atmospheric CO_2.

Vladimir refills the generator; notice the safety goggles and gloves. *Credit* MDRS-5 crew

The goal of today's EVA expedition was to find two places: a first one that could be intermittently wet and a second exposed to wind gusts, in order to collect soil samples from both places and to start ecosystem cultures back at the lab. The sustaining idea for our biologist Nancy is to verify whether it would be possible for potential Martian bacteria-like organisms to be transported by Martian winds. It is true that the weather today is supporting that idea, as it is windy with gusts of more than 50 km/h and red desert dust flying all over.

Tonight, our DGO has promised us out-of-this-world fajitas so I will sign off here for this third day, as I do not want to miss that.

Evening regards from Ares.

On to Mars!

Vladimir
Cover e-mail

Hello everybody,

Felt great today. Nice and busy day. But still some connection problems with the computer. I can access the external ESA web page but I cannot access my e-mails at ESA. So, if you have sent me something, please have patience, I will answer eventually. If really urgent, please answer at this e-mail address. Be aware though that this is a common address shared by all the crew and all the mission support team in California (this makes a lot of people), so don't be too chatty and make sure you put my name in the subject field.

Photos are coming in a parallel e-mail.

Best regards to all.

Vladimir
Wednesday 10 April 2002, Day 4

Martian greetings, Earthlings!

What an extraordinary day we had today! There was a big excitement this morning in our Mars Desert Research Station in the desert of Utah. Less than 48 h after being planted, our seeds started to sprout! The miracle of life took place again! It is extraordinary! Of course, it happens every day all around the world: watering a seed in the ground will eventually make it sprout. But here in our lab in this close isolated research Hab environment, it looks extraordinary. In fact, all the four kinds of seeds, the alfalfa, the tatsoi sprouts, the arugula salad and the radishes, planted in a rock wool tray installed in the lab started to came out overnight.

First sprouts (middle bottom part) of seeds in the rock wool tray in the Hab. Great! *Credit* MDRS-5 crew

Nothing is visible yet from the seeds installed in potting soil trays, most probably because they are deeper in the soil. I will keep you informed of their growth and of the others when they come up to life. We hope to be able to consume them before the end of our rotation, in about ten days.

This morning, my first real (nearly) warm shower since a few days, which brings you back into mankind.

Later on, I eventually succeeded in getting through to my e-mail system at work, miracle of modern technology. The culprit for taking so long was a bad connection in the Hab that took us two days to track. The only computer that is not yet on line is Andrea's, everybody else is up and sailing on the Web.

Andrea using the Hab computer to check her mails. *Credit* MDRS-5 crew

I started to go through my mail and I found a video recording sent by my 13 years old son. Unbelievable how today's youngsters manage to fiddle with electronic gadgets, in this case a new web cam.

A video message from outer space. *Credit* VP

Everything is fine at home and he would like to know how he could help us. I let him know that it would be nice if he could search on the Web for some information on how astronauts and cosmonauts do to clean up the space station, what are the respective tasks to compare with our daily chores.

Later on, with Nancy, we sorted out also the Eco data logger that we installed back in the Hab in a semi-EVA mode, that is pretending to wear only oxygen helmets.

Nancy showing the Data Ecologger in the Greenhouse. *Credit* MDRS-5 crew

After the last two days' exuberances, our Commander, the DGO of the day, reintroduced some rigor in the lunch approach and we went back to self-made ham and cheese sandwiches. Not that anybody complained.

The purpose of today's EVA was to rediscover a GPS waypoint set up by a previous crew for geology investigation purposes. Some other tasks needed also to be done during this EVA. Andrea, the mission geologist, and I were assigned for this EVA. As we were leaving by ATV, the wind picked up again to become nearly stormy. The Martian flag (recall: blue, red, green) had to be brought back in the Hab. We took a few GPS waypoint coordinates around the Hab, and we set our course to *Candor Chasma*. This is an extraordinary place!

In fact, the Hab is surrounded by desert. But this desert is not uniform and boring. No, to the contrary: it is rich in colours and full of dramatic landscapes as the pictures show. Crossing some plains with only red dust and rocks, we arrived at *Candor Chasma*. This place is named after a real feature, a huge canyon, on the planet Mars and here on Earth, it is a canyon about 50 m deep with a small hill in the middle divided by horizontal geological layers of different colours.

Left: Candor Chasma in the Utah Desert. Right: Vladimir measures the GPS coordinates. *Credit* MDRS-5 crew

Left: Knob Hill, close to Candor Chasma. Right: Red dust and rocks. *Credit* MDRS-5 crew

Many of these landscapes can be found around here and are so similar to what we know of the Martian surface that this is one of the best places to conduct Mars mission simulations. We could not locate with precision the waypoint that we were searching for. We were blocked by an impassable obstacle some 800 m away from the waypoint. We probably have to take another route. As the storm was getting really bad, we were ordered to return to the Hab and while backtracking, we managed to get blocked in the sand, me first, then Andrea.

Left: Andrea driving in the red sandy dust. Right: Andrea got her ATV stuck in the sand. *Credit* MDRS-5 crew

But we made it eventually just in time before the hard rain and the thunderstorm. Quite impressive, a real thunderstorm in the desert. The air was so full of static electricity that we were treated also to a Saint Elmo's fire in the Hab close to the lightning rod. All this while we were eating our dinner cooked by our DGO Commander (spaghetti and beans salad). Yes, it was a good day! And to stay in the Martian mood, we watched tonight "Red Planet" (terrible). Fortunately, enough, it does not describe at all the way Mars will be conquered. Till then, good night from Mars in Utah and with Martian regards to all.

On to Mars!

Vladimir
Cover e-mail

Hello everybody,

Great today, but again much to do. Our server via the satellite went down again at 1 h 30 this morning. So, it is late again, passed 2 h in the morning. So, I will be brief. I ran out of battery for the digital camera and I do not have a charger. I will rely on someone else's camera tomorrow, hoping that a charger will become available. I could eventually connect with the e-mail system at ESTEC so every-thing is back in order and I will start to answer all e-mails in the coming days.

Photos are coming in a parallel e-mail.

Best regards to all.

Vladimir
Excerpt from the daily activity logbook

Great EVA today with Andrea. Came home at around 20 h, and dinner was late. I manage to start writing my daily reports in between. We watched a stupid DVD from 10 h 30 till 12 h 15. Then I tried to finish my reports. I went to fill up the generator with Jan at around 1 h 15 in the morning. Then, I finished my reports at around 2 a.m. Tried to connect to the e-mail system at the office to send the reports and the photos. Extremely slow, which is not surprising as my connection is in competition with all the e-mail traffic in Europe. My e-mail with the texts passed through at around 2 h 30. The e-mail with the photos bounced back and came back 45 min later and I resent it. I decided not to wait any longer and I crash-landed in bed at 3 h 30. I woke-up at 6 h 15 to go to the bathroom, and fell asleep again to wake up at 8 h. Stayed in bed till 8 h 45 for a quick breakfast and the nine o'clock briefing. I write this at 11 h 40. Not yet had the time to wash and brush my teeth. Time is flying too fast and there is too much to do. But it is still great and fascinating.

Thursday 11 April 2002, Day 5

Martian greetings, Earthlings!
We have now established our normal sailing speed and life continues normally with its lot of joy, discoveries and excitement. We have the pleasure to announce to you the appearance of 22 new little sprouts in the potting soil tray in our living room and their big brothers and sisters that we observed yesterday are doing fine as well (the longest is already longer than 10 mm). But it seems that their cousins that we have installed in the greenhouse are not as fast: we observed only a few coming close to the surface.

The first sprouts coming out in the potting soil tray in the Hab living room (left) and in the rock wool mat in the Greenhouse. *Credit* MDRS-5 crew

These last few days we discussed also Yuri's night that will take place on Friday 12 April for the anniversary of the launch of the first Earth cosmonaut. There will be parties all over the world, literally as there will be a celebration also on board the space shuttle and in other "space" places. We intend to also have our own celebration on this occasion and you are invited to join in, only in mind and spirit unfortunately, to celebrate the accomplishment of human beings in space for 41 years.

Today, I would like to tell you more about our daily life in our Mars Desert Research Station. First, let's do a quick tour of the location. The Hab is a cylindrical structure 8 m in diameter and 6 m in height divided in two floors. The ground floor has two airlocks, at the Hab front and back, an EVA preparation room, a large laboratory area subdivided in biology and geology areas, a small shower/sink room and… a toilet (yes, you guessed it right). One accesses the upper floor by a steep stair. One half of the circular floor is occupied by a common living area with a kitchen corner, a semi-circular workbench were all computers are installed, a table for six persons where we have our meals together or where we hold our briefings and meetings. The other half is divided into six small bedrooms where one room sleeping bunk is superposed to the one of the next room. Additional storage place is found above the bedrooms, with a circular hatch in the centre of the ceiling. Two circular windows look to the East and the South. Smaller windows give additional daylight.

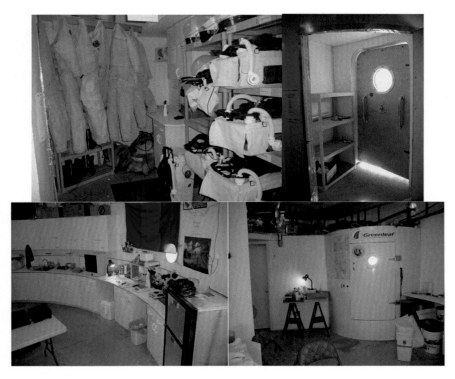

Views of the Hab on the ground floor: top left: the main airlock; top right, the EVA preparation room; bottom left: the lab for biology experiments; bottom left: the second back airlock and the bathroom door on the left. *Credit* MDRS-5 crew

This is the home of our crew for two weeks and I must say that it is relatively comfortable and spacious enough. We are generally so busy that we do not bump into each other and in fact, there is no personal space problems.

Where do we work the most? Top left: Andrea works on the MDRS computer; top: Bill (middle) and David (right) work in their respective rooms; bottom left: Nancy works on her computer on the circular table in the living room; Jan cleans the dishes in the kitchen corner as the day's DGO. *Credit* MDRS-5 crew

We are participating also in some long-term experiments like the assessment of the quantity of water and soap we consume. Drinking and cooking water is not restricted but we are asked to refrain from using too much water for cleaning and washing. I enjoyed yesterday morning my first thirty seconds' shower and it is in moments like these that you really appreciate this comfort. We were also delivered 80 grams of special NASA soap for the two weeks. This soap is odourless, tasteless but foams like hell even for the few milligrams you could take on the tip of your fingers.

There were two EVAs today. The first one allowed Nancy to install some collecting apparatus to collect flying dust in the hope to find airborne living organisms. For the second EVA, Bill, David and I went with the ATVs searching for two waypoints. *En route*, we found a geodetic point installed by the US Department of Geodetic Survey and corresponding to a round GPS coordinate. We found also a breath-taking landscape full of canyons, hills, so colourful and extending as far as you could see, results of eons of erosion and geological rock crushing. So, bare and desolate, you could expect to see a dinosaur popping out anytime from these ancient layers.

Breath-taking landscape with Henry Hills in the far and Skyline Rim to the right. *Credit* MDRS-5 crew

And so, life goes on in this extraordinary environment, shared between conducting experiments and observations, writing reports, going on EVA expeditions and discussing space and Mars exploration over meals. What a place for scientists to meet! Nancy just showed us with the microscope a still unidentified living organism trapped in some crystallised structure, that she collected in some rocks.

Nancy working in the biology lab on the ground floor. *Credit* MDRS-5 crew

This evening at sunset, as the sky was incredibly clear, I showed to the rest of the crew from the roof hatch where and how to observe the visible planets with a pair of binoculars. We could clearly see Venus and its phase, Jupiter and some of its satellites, Saturn, and a double star. Simply great!

Jan is our DGO today and he is preparing us *Candor Chasma* chicken sauté and Wells war of the hash browns. What else would you expect in such a scientist's paradise!

Reach for the stars! They are nor far!

On to Mars!

Vladimir
Cover e-mail

Hello everybody,
Nice day again today, but a bit tired so I will try to go bed earlier than these last few days.
Still no battery for the digital camera and relying on my Commander's camera. So far so good, but I would need my own.
Photos are coming in a parallel e-mail.
Best regards to all.

Vladimir
Excerpt from the daily activity logbook

Great EVA outing this afternoon with Bill and David to find two GPS waypoints. Of course, did not find them. Nevertheless, visited a geodesic point and a fascinating landscape. Continued till the waypoint WP14 and found cows and cowboys. Decided to turn back. Maybe went too far … Back in the Hab, observed planets at sunset. Super cool!
Decided to go to bed early, but was still up till 2 a.m. to write and send daily reports. Went to refill the generator with Jan at 1 a.m. to replace Andrea who is already asleep. Bed at 2 a.m. and woke up twice to go to bathroom. Up at 8 h 30. The generator fuse blew up and went to put it back in place with Andrea before breakfast.

Friday 12 April 2002, Day 6

Martian greetings, Earthlings!
Today is a special day. Some decades ago, an intrepid young man braved the laws of Nature and embarked on the first extra-terrestrial journey. We are all somehow following his footsteps and preparing for mankind to expand his vision, knowledge and presence on other worlds.
Tonight, we celebrated Yuri's night and we had a toast to him and other space travelers who gave their life in opening this new frontier, to all the accomplishments of cosmonauts and astronauts since 41 years and to the future of mankind on other planets.

Toast for Yuri's night; from right to left: bottom: Bill, Andrea, David; top: Jan, Nancy and Vladimir. *Credit* MDRS-5 crew

We had a special meal this evening, composed and produced by today's DGO Nancy. Special in several senses, as Nancy used for the first time some herbs harvested in our greenhouse, thyme, lavender, and chives, to accompany our feast dinner made of *Olympus Mons daube de boeuf.*

Left: Lunch and briefing, from left to right: David, Vladimir, Jan, Andrea, and DGO Nancy. Right: DGO Nancy preparing the evening dinner. *Credit* MDRS-5 crew

Evenings like these are unique among people sharing the same interest and enthusiasm for space exploration. We watched also a spacey movie and our choice fell on "2001: A Space Odyssey" as being the most appropriate for this special evening.

Well, the rest of the day went along well. The weather was cloudy today and relatively chilly, around 15 °C (59 °F), and the forecast is for worse weather in a few days (someone even mentioned the words sleet and snow …). We will see.

Good news from our various plants. The radishes and alfalfa are growing fast and well in our Hab living room. Last night, we sampled a very few of them among those that clustered too close to each other to grow to full maturity. The other plants are growing nicely as well in the lab downstairs. But surprisingly, their cousins in the greenhouse are not doing so well. Most likely environmental conditions of temperature and humidity are not as optimal as in the Hab.

A view of the indoor plants on the fifth day. *Credit* MDRS-5 crew

We had also some technical surprises. Some good and some bad. The good one is that Andrea's computer is now connected and working. One of the bad ones was this morning, when tragedy struck at breakfast: one of the fuses blew on the generator and the two persons on duties, that is Andrea and me, had to go and reset it while still half asleep (that was the tragedy!). And then, the biolet got stuck ("*Roses are red, Biolets are blue,…*" was singing Nancy). But you don't want to hear that story. It already happened last year in the Arctic, but it is still unpleasant. Luckily, our Health and Safety Officer Jan Osburg sorted the things out with a hammer and spanner.

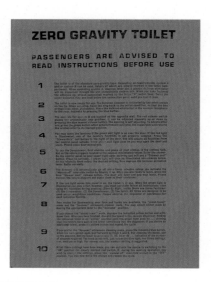

No comment. *Credit* MGM/MDRS-5 crew

On a better tone, today's EVA was again multipurpose. Commander Bill Clancey, geologist Andrea Fori, and journalist David Real took the ATVs first, to retrieve the airborne sample collection devices installed yesterday by our Biologist Nancy Wood; second, to take photo documentation of interesting geological locations; and third to observe fossilized oysters in an ancient seabed from the Cretaceous period, 140 million years ago. Many photos were taken and are now placed on the website of *The Mars Society*'s Desert Research Station.

Left: Colourful landscape. Right: Andrea looks at Henry's Hills. *Credit* MDRS-5 crew

Left: David and Andrea climbing Oyster Hill. Right: At the top of Oyster Hill. *Credit* MDRS-5 crew

As I was not involved in this EVA, I could catch up with some reports and I prepared a Science summary report on biology, astronomy and geophysics activities for these first four days of our mission. So, a quiet day before the evening's party.

Signing off from a peaceful desert spot reminding us of future Mars,

With Martian greetings

On to Mars!

Vladimir
Cover e-mail

Quiet day today, but a great party ahead of us. So, I will try to finish this a.s.a. p. Still without battery for the digital camera and still relying on Bill's camera. I may not write tomorrow, as it is Sunday. It will depend on my mood and courage tomorrow. In parallel, photos are following.

Best regards to all.

Vladimir
Excerpt from the daily activity logbook

Yuri's day and night. No EVA outing for me today. I stayed in the Hab to finish my Science report. We toasted a first beer to Yuri at 19 h 45, then a great dinner thanks to Nancy with wine for the occasion. Succeeded in sending all reports at 22 h 30, while we were eating. Then, watched the DVD "Outland" with Sean Connery (an excellent classic) from 11 h till 12 h 40. The other DVD, "2001: A Space Odyssey", could not be found.

Around 1 a.m., last refill of the generator with Andrea and last refill of the water container on top of the Hab. The water pump broke down and we started thinking about alternative solution. Dismantled it with Jan and saw that it was irreparable. Finally, in bed at 2 h 30. Woke up twice at 4 h and 6 h. Eventually got up at 8 h 15 this Saturday morning.

Saturday 13 and Sunday 14 April 2002, Days 7 and 8

Martian greetings, Earthlings!

The first part of this weekend was more than interesting. Saturday morning, we ran into a mini-crisis as we realized that we would be short of water before the end of the weekend. Let me explain. We still depend on external sources for two main supplies: water and fuel. Both are brought from the nearest village (approx. 20 km from the Hab) by a local provider. To order any of these two, we send the order by e-mail to Mission Control in California, who then contacts that person. Fuel was replaced on Thursday; water not. So, we were contemplating to stay without water throughout the weekend. Saturday morning, the entire crew came down to try to lift this cylindrical tank of 3 m diameter and 1 m high to siphon the last litres. Not easy. My idea to lift the tank with five persons did not work but Bill's idea to roll it on its side did eventually work with some modifications. We could eventually siphon the remaining litres of water from this tank of 2.5 tons (when full) that we are depending on. We started to envision measures like no more washing, and no more dish washing. We had also another problem: the night before, the little pump used to pump the water from the big tank to the top of the Hab broke down unexpectedly. So, we had no means to pump the remaining water from a nearly empty tank. The situation looked bleak.

Nevertheless, as we are explorers, we decided to not deviate from our plan of action. We had set our mind the day before on an ambitious EVA expedition to explore and revisit ancient points visited by previous crews. And it was what we were going to do! With Jan and David, we planned our expedition very carefully by studying the map and recording all previous points on our GPS's. We put on our EVA suits after a light meal to leave in the early afternoon. I was the EVA Commander and on our ATVs, we set our bearings to North, and then turned West in the hope to find a mud track indicated on the map. We never found it. But we were taken through extraordinary landscapes and hills and canyons. We missed the first GPS waypoint, but we found all the others. Some of the paths that we took reminded us of those followed by Luke Skywalker in "Star Wars", either following ancient river beds at the bottom of multi-layered canyons or going up and down hills close to canyon edges. We found also an unsuspected field of fossilised oysters from the Cretaceous, like our colleagues did a few days ago, in another area. It was unbelievable. But this was only the first part of our trip. Once we came back on the "main road" (a dirt road called the *Lowell Highway*), our next goal was to explore further on and to find a new way to the river. *En route*, we took several GPS positions and we named in turn several features on the road, like for example *Dimitri's corner* (on *Lowell Highway*), or *Brussels sprout* (a hill in the middle of the road that I climbed), etc. We eventually found the river, called *Muddy Creek* river, coming to the edge of a cliff from which we had a vantage view out of this world and most likely the same that dinosaurs enjoyed many millions of years ago. From there, we backtracked and we set course on our third objective, which was to locate precisely a water feature indicated on the map. We found it but it was not surface water, it was an underground reservoir that we could eventually spot because of the vegetation it harboured, a sort of oasis of a few trees and bushes in the middle of rocks and sand. Our hypothesis was quickly confirmed by the sign of the US Dept. of Geology Survey indicating this natural underground reservoir. We collected some soil samples from that area for our biologist, and we eventually came back to the Hab, after a four-and-a-half-hour ride in the desert. What an unforgettable experience!

En route for a great EVA expedition. Top: left: measuring position with respect to recorded Way Points; right: stopping at a US Geological Survey landmark. Middle: left: at site called "UFO landing site"; right: at *Dimitri's* corner. Bottom: left: at *Brussels' sprout* site; right: at *Muddy Creek* river. *Credit* MDRS-5 crew

In the meantime, the deliveryman came to fetch the tank and returned it filled. So, the only problem remaining was the pump. Not being able to fix it on the spot, we decided to carry the water by hand and the six of us forming a chain, we mounted bucket after bucket 250 litres of water. What an exhausting day!

Luckily, this Sunday morning, our Commander decided to grant us "individual work schedule", which is a very nice way to say that we were free to do what we wanted. So, I could enjoy my first late morning and a sponge bath. Yes, due to the new water situation caused by the pump problem, 30s' showers are not allowed anymore. A sponge bath is not so bad after all, and you would not believe how much you could wash with only a glass of cold water and a sponge. As today's DGO, I fixed us scrambled eggs and bacon for the brunch and we staged a plot for the afternoon. After all, it was Sunday afternoon and it was time to test Mission

Support abilities to help a stranded Martian crew. The plot was simple: two simulated EVAs would take place simultaneously, one with three persons to retrieve biological samples, and the second with two persons in the greenhouse. But then everything would start to go wrong. The first EVA team would be separated and lost in a sandstorm while the second EVA would stay stuck in the greenhouse due to a blocked door zipper. On top of that, a general power failure would force the remaining person in the Hab (role played by the Commander) to leave the radio station and go to fix the generator, leaving Mission Support in charge of remotely helping everybody. Well, we had a good laugh during this simulated simulation and Mission Support eventually guessed that something fishy was going on. This Martian joke was in fact a good training session for Mission Support to realize where the problems were in the overall mission scenario and organisation.

The news from our Garden of Eden are excellent as the radishes are growing seemingly at an exponential speed. Saturday, the tallest radish leaves went from 4.4 cm on Saturday to an astounding 7 cm this Sunday evening.

Unbelievable 7 cm tall stems of radishes in the Hab growth tray. *Credit* MDRS-5 crew

For the evening, I have prepared *Schaeberle* chicken sauté with *Deimos* rice salad, enjoyed by all present Martians. And to conclude this day, we watched another story of a world without water, the first part of the remake of "Dune" (not bad so far).

From a planet full of diversity and life, I wish you a peaceful Martian night on Earth.

On to Mars!

Vladimir

Hello everybody,

A nice refreshing weekend, full of adventure on Saturday and quiet on Sunday. Still no battery for the digital camera. Texts and photos are following.

Martian regards to all.

Vladimir
Excerpt from the daily activity logbook

Saturday morning, woke up at 8 h 15. Without washing and tooth brushing and before breakfast, went to see if possible to implement another solution to empty the tank. Came back at 9 h 00. Got some breakfast until about 9 h 40 and all crew went out to raise the water tank. My initial solution did not work but Bill's idea to tilt it was OK. Started to empty the tank by siphoning and around 10 h 30 left Nancy and Andrea to do the job, to prepare EVA navigation with Jan and David, until lunch at 12 h 30. After lunch, started to prep up at around 13 h 45 and left airlock at 14 h 15. Went to put garbage out and hit the road with ATVs at 14 h 30 until 18 h 55. Came out of airlock at 19 h 00.

No daily report today, but will write one tomorrow. Spent evening writing other things and again trying to send this Science Report. Took me quite a good part of the evening to reduce the size of the photo files and then finally to send it as a PDF file. Around midnight, a car arrived with three youngsters, obviously drunken. They turned around the Hab and with Jan we were ready to intervene, but they left by themselves around 12 h 40 (…).

Monday 15 April 2002, Day 9

Martian greetings, Earthlings!
At the moment I am writing (noon on Monday), we are in the middle of a severe Martian dust storm with winds blowing up to 80 km/h. That does not sound like a lot, but when you are in the middle of a sand desert, you take it right in the face. Literally. We can neither see the sky, nor the sun anymore, just dust everywhere and all around. We just had to consolidate the South window, as it is the one taking the full blast from the Southern wind. The whole Hab is creaking and swaying, and we wonder whether it will hold the blast.

Dust storm seen from the Hab window. *Credit* MDRS-5 crew

This morning, before the storm, I went checking the greenhouse and we are running low on irrigation water there. Our plants are doing well. The tallest radish stem has now grown to 8 cm from an already astounding 7 cm less than 12 h ago. We sampled two radishes and although we did not see any red bulb, we tasted the stem and leaves with Andrea and it tasted delicious. In the lab, I have checked with Nancy our other plants and we also sampled and tasted the arugula salad. Delicious!

Right: Andrea fills in the plant growth sheets; left: Vladimir measures the radish stems. *Credit* MDRS-5 crew

We just had lunch. David, our DGO, prepared sandstorm soup (just drop the black pepper grinder in the pot of soup and you'll understand!) and red-hot tuna tacos. David is from Texas, as you could guess. We are still planning to go on EVA this afternoon but after checking the weather map received by satellite, we would be better off waiting as a real thunderstorm with strong winds is announced.

It is now 7 p.m. Well, we eventually went on an EVA. We planned initially two EVAs: one close to the Hab for bio samples retrieval, the other exploratory with the ATVs. Nancy Wood, our biologist, and I went for the first one, to collect samples of soil in the immediate vicinity of the Hab to assess the bacterial contamination that human activities introduce in the environment. Six samples were supposed to be collected. So, no big deal, just get on your knees and fill up those little vials. Right, but not with an EVA suit of 20 kg on your back and not with a sandstorm blowing at 80 km/h. We tried a first time, but the plastic bag holding the empty vials got blown away, and don't even think about running with an EVA suit on. So, we decided to abort this first attempt and to return to base. We reconfigured our suits and our pocket content and we tethered absolutely everything. We went back in the inferno and it did eventually work, but it took us longer, about twice as long as anticipated. But definitely a good training and simulation of EVA activities in a Martian dust storm. After all, on Mars, dust storms can last weeks or even months and the first crews would have to face situations like these during their EVA outings.

Left: the Hab in the sandstorm with Vladimir in front. Right: with Nancy trying to collect soil samples during the storm. *Credit* MDRS-5 crew

After coming back to the Hab, we were welcomed by a strange and vaguely familiar smell. No, this time it was not the toilet, but freshly baked bread. Yes, our DGO David got the bread-making machine brought by Andrea eventually to work and we have a fresh and still steaming bread loaf on the table.

We have now passed the one-week mark of living and working together and we are complementing each other quite well. There is no tension among the crew and we have quite a good laugh regularly. Today could have been difficult, with this crazy wind-blowing non-stop, but it does not seem to affect the crew too much. We had Elton John and Bruce Springsteen (DGO's prerogative) on the loud speaker and the swinging in the rhythm of the wind gusts must have helped, I suppose. And here comes the rain! It just started and will help to wash out the dust accumulating everywhere.

Working in the Hab: left: after the EVA, each of us occupying a different place in the living room (from left: Nancy, Vladimir, Andrea, Jan); right: view from the top of the circular table (from left: Jan, Andrea). *Credit* MDRS-5 crew

Ophir Chasma pork medallions are on the menu this evening, with dusty planetesimal-shaped bread, red Martian green beans and fajitas. And we are looking forward to watch the second part of "Dune" this evening. So, all in all, everything is well under the Martian storm and on board our shaking and rocking ship.

Wishing you all the Martian best,

On to Mars!

Vladimir

Cover e-mail

Hello everybody,

A very stormy and dusty day, but all is cool on Mars. Photos are following hereafter.

Martian regards to all.

Vladimir

Tuesday 16 April 2002, Day 10

Martian greetings, Earthlings!

After yesterday's sand storm, we had a much cooler and nicer weather today. Cooler because the temperature dropped down to 15 °C (59 °F), and nice because it was blue sky and sunshine. I overslept this morning and got up at 9 h just in time to catch the morning briefing, missing breakfast. Just to show that we are quite busy here and that again I finished my reports and did some additional work for my job at ESA over the Internet until late last night. The morning was spent again finishing and correcting previous EVA and science reports.

Considering the weather change, we decided on an exploratory EVA today where Andrea, Jan and I would try to retrieve some bio sample collectors at various locations and to explore a new way to get to the river to collect mud samples for our biologist. After a quick lunch prepared by our Commander Bill Clancey, we suited up and I managed to find a new way to suit up completely alone without any help. We left the Hab at three o'clock and after retrieving the first dust collector near the Hab and re-erecting the Martian flag, we left on our ATVs in a northerly direction on the *Lowell Highway*. I was again EVA Commander. We measured several points *en route* using our very handy GPSs, but still managed to somehow have different readings. *En route*, our geologist Andrea took several photos of interesting features, mainly rocks and various geological layers. We passed *Dimitri's corner* and the *Brussels sprout* marks, to go further to the *Route 66* mark. From there we turned to a westerly direction to try a new route to the river. The main dirt road quickly ended in several paths among the rocks that we explored one by one, and taking GPS coordinates each time we could not go any further because of a dead end or a cliff. We decided to backtrack to *Brussels sprout* to try a different road from there, but to no avail. We backtracked again and went back to *Dimitri's corner* to take the other dirt road leading to the geodetic point further up West. This is typical of exploration expeditions where you see on the map where you want to arrive, but you do not know how to get there as there is no existing route and you often come against features that you cannot cross.

Two Martian-like landscapes encountered on our way during the EVA. *Credit* MDRS-5 crew

From the geodetic point, we turned North and followed the main canyon, first following a path on top of the canyon, then gradually coming down by different ways and eventually arriving in the canyon itself. We followed the canyon road that reminded us the other day of the place where young Luke Skywalker met Obi Wan. We continued on the canyon road and passed the point where we could not proceed any further previously. We called that point the Y Ravine junction, as two ravines come together making it nearly impossible to pass. Well, we managed to find a way to pass eventually to continue among further dramatic landscapes.

Left: Andrea takes a bearing with her GPS in front of Vladimir. Right: Andrea and Jan at the Ravine Junction before crossing. *Credit* MDRS-5 crew

We found the access to the river at the end of that canyon and we ended up on a sort of beach. Happy and relieved to have found it, we collected the two mud samples and we decided to cross the river and to explore its other side.

Andrea and Vladimir collecting mud samples from the river. *Credit* MDRS-5 crew

We drove upstream to soon arrive in front of rock formations that we could not climb with the ATVs. We had to cross again the river. Leading this expedition, I tried to pass first and, although the first few meters went fine, I soon realize why it was called the *Muddy Creek* River: the rear wheels of my ATV got sucked in the mud and started to sink deeper and deeper. Yes, like in the Arctic. I could no longer go forward or backward and the engine stopped. We quickly devised a rescue plan by radio with Jan and Andrea for them to go back where we crossed initially, to cross back and to meet on the other side. Luckily, we took a long rope that Jan had on his ATV. Attaching the rope to both ATVs allowed Jan to tow me out of that mud trap. Once safely on a sort of little island in the middle of the river, I could restart the engine and finish the crossing.

Left: the ATV of Vladimir stuck in the *Muddy Creek* river. Right: the ATV being towed out of the mud by Jan's ATV. *Credit* MDRS-5 crew

This rescue having eaten most of our remaining time, we decided to drive back to the Hab by the shortest way, as the night would fall in about one hour. On the way back, the three of us got stuck in turn at the Y ravine junction. It took the three of us to pull and push together each ATV out of the soft sand ravine. At some point, Andrea fell on her back, lying down stuck on her backpack, like a tortoise. Impossible for her to stand up or to roll back on her side alone. We left our ATVs, Jan and I, and we ran to help her, laughing and happy that nothing serious happened to her. We made it eventually back to the Hab, after having retrieved on our way two other dust collectors. Again, what a great ride it was! A five hours' expedition in the desert and in these beautiful Martian-like landscapes! And also, many lessons learned with relevance to Mars exploration, e.g. a stranded crew that manages by itself to rescue one of its members.

We arrived at the Hab just in time for dinner as our Commander, fulfilling his DGO duties, had prepared us space meatballs with pasta and *Valles Marineris* sauce and freshly made bread. To finish this great day, we watched the last episode of "Dune" (really a good film) until around midnight.

Have a repairing Martian night on Earth.

On to Mars!

Vladimir
Cover e-mail

Hello everybody,

A very nice day, and all is still superb on Mars. The photos are gone in a parallel universe and are coming back to you.

Martian regards to all.

Vladimir
Wednesday 17 April 2002, Day 11

Martian greetings, Earthlings!

Today we had still to battle against strong winds from the South. Our ship is again rocking and swinging against the elements. Our greenhouse was nearly blown away and needed to be secured and repaired. So, we spent a good part of the morning with Jan to secure the cargo straps around the greenhouse structure and to retighten the ropes holding it attached to the ground. The door of the greenhouse was ripped apart and blown away by wind gusts more severe than two days ago. We taped it with duct tape the best we could but I am not sure it will hold the night.

Today the EVA expedition concerned biology and geology field exploration. Our Commander Bill Clancey took geologist Andrea Fori and biologist Nancy Wood for an explorative EVA in *Lithe canyon* where rumour has it that there are dinosaur bones hidden in the ground. Jan, David and myself are on duty to guard the Hab and to avoid that it gets blown away or sunk by the wind and the sand.

The EVA party Bill, Andrea and Nancy in *Lithe canyon. Credit* MDRS-5 crew

I spent nearly all day writing and polishing reports, fixing what could be fixed on the greenhouse and thinking how much this simulation is different and how much it is similar to the one of last year in the Canadian Arctic Circle.

First, both Habs are very similar but there are certain differences: the stairs from the ground to the upper floor are steeper and placed in the other direction, which makes the ground floor bigger and provides more room for the biology and geology labs; the individual rooms are fitted with working space and shelves but not sufficiently. Second, the way we are scheduling our activities is different. We have here two weeks to accomplish our research work compared to only one-week last year in the Arctic. This means that we are more relaxed on the work organization, but not meaning that we have less to do (on the contrary...). We are also more involved in discussing daily schedules and decisions. Improvements were made also on the EVA equipment and suits. For example, we have now portable rear view mirrors that can be strapped to the arm or the glove, something missing last year and that several crewmembers complained about while riding ATVs. Also, the greenhouse was introduced. Of course, it is not yet fully operational, but we are using it already for some plant growing and some biology experiments. It will become a major subject of research in the coming years to see how crews can grow part of their food supply by themselves to better prepare Mars mission scenarios. Finally, the isolation is different. Last year, we were in a more remote area and more difficult to reach on Devon Island; but the proximity of the base camp, radio communications between the Hab and base camp, and the quasi-constant presence of a *Discovery Channel* cameraman and from time to time of a journalist rendered this isolation feeling more relative. This time, although we are in the middle of the desert, we are only a few kilometres away from a little village in the middle of the USA. However, the absence of radio contacts and the delay in e-mail communications, the absence of any visitor (except for the guy who brings the water and the fuel every other day and with whom we have no exchanges) render this isolation more real. We are really living only the six of us and we can only talk to each other, depend only on each other, interact and communicate only among each other. And it works well. We did not have any crisis or interpersonal problem. Everything was faced and solved with efficiency and in a good atmosphere.

There are certainly things to be changed and improved, but gradually all comments and lessons learned are incorporated in the future simulation scenarios and in the design of future Habs. The next one to be built will be installed in Europe in 2003 or later in Iceland. This Hab is near completion, to be exposed in Chicago from June to October 2002 and then will be shipped to Europe. The fourth Hab will be installed in the Australian outback desert sometime in the future.[2]

Tonight, we will go out and watch the sky for the planet and Moon alignment. On to Mars!

---

[2]As said, unfortunately, these Habs were never deployed for budgetary reasons.

Vladimir
Cover e-mail

Hello everybody,

Still strong winds and greenhouse nearly gone, but we're on track. Photos are coming back from the neighbour parallel universe.

Martian regards to all.

Vladimir
Engineering Report to Mission Control on the state of the greenhouse, 17 April 2002.

This morning, we have inspected the greenhouse with Hab Chief Engineer Jan Osburg after the strong winds we had these last two days. Actually, this morning the wind was still blowing quite strongly.

We have made the following observations and tried to repair as much as possible.

(1) The structure still seems OK. We tighten the cargo straps around the cylindrical surface and the six ropes holding the base to the ground. We put duct tape on the hooks holding the cargo straps to avoid them coming loose under the wind.

(2) There are damages to the material that makes the cylindrical wall of the greenhouse. We could see three places where the wind tore apart that material (on top and on both sides). It seems useless to repair that with duct tape: first, it is difficult to access with the wind still blowing; second, the wind will immediately rip it apart again. So we left it as it is, hoping that the next technical crew could fix it more permanently.

(3) The zipper door that was placed as repair on Sunday 7 April by Frank Schubert and his team was blown away. The duct tape holding the blue sheet were blown away first, then at lunch time today, the zipper completely disappeared, blown away by the combined action of the wind and the sand. Then, the wind ripped apart the replacing blue sheet. We taped it back as much as we could, but this definitely needs a more permanent fix/repair, hopefully by the next technical crew.

The greenhouse door was ripped apart by the wind and fixed by Jan with tapes. *Credit* MDRS-5 crew

(4) I have added two more buckets of water to the blue tank where the pump is still active. The feeding hoses that lead to the distribution racks came loose several times from their attachments on the irrigation rack and were put in place but again blown away by the wind.

I have attached some photos for you to better visualize the problems that we have. The most damaging causes are the strong gusts of winds that pull everything away.

In general, I think that the design of this greenhouse was not conceived to support winds like we have here, and should be reassessed for utilization on Mars.

For example, if a cylindrical shape has to be kept for pressurization, why not put the greenhouse on one of its circular sides and keep a vertical cylinder architecture? You end up with a greenhouse where you can put more racks, a lesser height and less surface to the wind.

Let us know if there is anything else that we should do.

Vladimir
Thursday 18 April 2002, Day 12

Martian greetings, Earthlings!

Big excitement yesterday evening: Nancy discovered a scorpion under her bed. After verification on the Web, it was a *Centruroides Exilicauda*. Venomous and lethal for children. Amazing! Such a little animal, only 2.5 cm (an inch) long, and so unfriendly. We were reminded to verify our shoes in the morning as these charming creatures like dark warm humid spots.

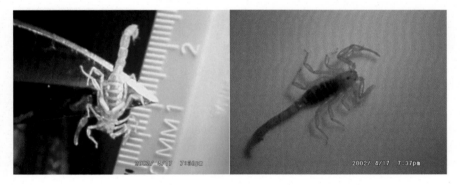

The scorpion *Centruroides Exilicauda* discovered by Nancy under her bed. *Credit* MDRS-5 crew

The weather turned cold during the night. The coldest temperature was 5 °C (41 °F). This morning during our briefing, we discussed the coming end of our simulation. It is true that time has passed so quickly and that we were so busy, that nobody really realized that we are nearly at the end. In three days, we will be leaving. In fact, the isolation will stop on Saturday already, as we will have the visit of nine media representatives, European and US television and newspaper journalists. After 'isolation', it will be 'invasion'. So, we started to prepare with the help of David, our resident journalist, what to say and how to say it. Tomorrow will be a big clean-up day also. Our Hab Engineer Jan Osburg noticed that the biolet was not functioning optimally (an understatement…). So, he decided to give it a good cleaning before tomorrow. We had to open all doors and hatches and for once the blowing wind was more than welcome…

This afternoon, everybody had a lot of things to do and to finish as suddenly everybody realized that in a couple of days, it would be all over. Nobody really was interested in an EVA, except for Jan and myself. So, we suited up in the middle of the afternoon and did a few things in the greenhouse, like replacing the data logger and brushing up the solar panels (yes, that is also something that astronauts would have to do on Mars after each sand storm).

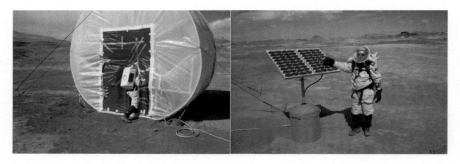

Vladimir entering in the repaired greenhouse in EVA mode (right) and dusting off the solar panels (left). *Credit* MDRS-5 crew

We proceeded then to *Candor Chasma* to revisit this splendid place that we visited on day 4 with Andrea. It seemed to me that it already changed after the sand storm and rain we had over the weekend. Could it be that landscapes on Mars are equally rearranged by wind like they are on Earth?

Vladimir in *Candor Chasma* canyon. *Credit* MDRS-5 crew

We left *Candor Chasma* to tackle the next item on our program, which was to find a way to climb *Skyline Rim*. Well, *Skyline Rim* is quite another affair, and a big one. It is a huge plateau made of cretaceous sandstone and rising nearly 130 m above the surrounding plain. Not easy to pass and not easy to climb with ATVs. So, we tried first toward the North without success, and then we went southward, to try to go around. And we drove, we drove, until we could not drive any farther as we came across a small river, but still no way in sight to climb *Skyline Rim*.

Two views of *Skyline Rim* to the left and *Factory Butte* to the right. On right picture: Vladimir leads the way. *Credit* MDRS-5 crew

We noticed as well that Jan's ATV had a flat tyre, so we decided that it was time to come back. We measured on our way coordinates of several geodetic points in the desert, so as to be able to relocate on the map our exact path. We came back safely to the Hab well in time for dinner.

In front of *Skyline Rim* (left) and taking a picture (right) from *Skyline Rim*. *Credit* MDRS-5 crew

Our plants are going well, thank you. They are growing like mad: the tallest radish stem is now 10.5 cm and a tatsoi stem in the lab downstairs caught up with it, at 10.5 cm, as well. Which one will win? We should know Saturday night when we will harvest them all and make a salad.

This evening, we will feast on *Nirgal* sate prepared by Nancy. Don't ask what it is. I think there are meat, unknown vegetables and some herbs from our Martian greenhouse. These are what make it so special.

This evening also we will have a break and watch a DVD. Discussions are on-going whether it would be "Matrix", "Spaceballs" or "Starship Troopers".

Well, you see, life is going on and the day is coming to an end with humans having the same preoccupations like in any other house on our planet, whether it is Earth or Mars.

On to Mars!

Vladimir
Cover e-mail

Hello everybody,
Realized today that it will be soon over. Still so much to do. The other parallel universe will deliver the photos.

Martian regards to all.

Vladimir
Friday 19 April 2002, Day 13

Martian greetings, Earthlings!

This is our last day of isolation. Tomorrow, we will have guests and visitors as we have an open house, where we will show anybody interested around. I will be the DGO tomorrow and I will have to find a way to feed 14 or 15 persons with whatever I will find in the cupboard, left-overs and tuna fish cans.

Today was a big cleaning up day. But only in the afternoon, as the morning was reserved for finishing up all work and reports. So, I watered the plants, talked to them nicely (as we will eat them tomorrow), measured them up (11 cm for the tallest radish stem in our living room).

Plants in growth trays in the Hab second floor (left) and in the greenhouse (right). *Credit* MDRS-5 crew

I finished a Science Summary report, the report of yesterday's EVA (both are accessible on the web) and some other things. I filled in a psychological survey questionnaire. And in the afternoon, I mastered the Shop-vac, the vacuum cleaner for industrial shops. It took about two and half hours but the place is clean now and there is no dust or sand around. You get used to having sand in your teeth, ears and eyes, but once in a while, it is nice to clean up the place. After that, Nancy and I went by foot to retrieve some of her vials that she placed in the sand not too far from the Hab. On our way, we found traces of a cougar, a mountain lion, and judging by the size of its trace, it was a big one. The desert is not so deserted after all.

Right: Cougar traces close to the Hab. Left: a plant growing in the desert; in absence of a better name, we called it Onion neurone plant. *Credit* MDRS-5 crew

And then, it was already the evening, a busy last day that we did not see go by.

This evening our chef-DGO Andrea prepared us a *Nectaris Fossae* pot with rice. Again, don't ask what it was. I do not know the secret. All I can say is that there was meat and vegetables, but which ones? All our supplies of fresh vegetables are gone since long. There was also some dried fruits and garlic, this I recognized. But it was good!

We watched a DVD together this evening also, it was the animation "Titan A. E.". Much better than "Starship Troopers" of last night, which was really bad.

This will probably be my last message to you from the Mars Desert Research Station. I will try to write you tomorrow but most likely I will only be able to send it on Sunday while in transit from Salt Lake City Interplanetary Airport on Earth. I enjoyed telling you all about what happened to us here while we were simulating this Martian mission for two weeks in the desert of Utah. I hope that it gave you some positive ideas about going to Mars as well. Hoping to meet you in this new step forward of mankind, it is a good bye from the Hab, a good bye from Mars, and a good bye from me.

On to Mars!

Vladimir
Cover e-mail

Hello everybody,

Last day of isolation. Nearly finished everything. Photos of the parallel universe are flying to you at warp speed.

Martian regards to all.

Vladimir

Chapter 6
The Desert After—The Last Day and the Return

Yes, the last day. What happened this Saturday 20 April 2002? Well, as announced, it was an extremely busy day. So busy that I did not have time to write anything. I will give you now some recollections of these last two days.

Upon rising Saturday morning, everybody was busy preparing for this 'open house' day and the visit of the medias. We all put on a badge with our name and position. Being the DGO of the day, after breakfast, I took upon myself to prepare some sandwiches and toasts for our visitors. As it was relatively cold and rainy, I also prepared two pots of warm soup and a few litres of coffee, all of this juggling with only two electrical hot plates, threatening to blow the generator fuses. The journalists and cameramen were foreseen around eleven a.m. And they arrived, thirty of them, with video and photo cameras, mikes, notepads and pens. There were American and European television stations: Fox TV from Phoenix in Arizona, a TV from California, RTL-TV from Germany, another TV from Sweden; journalists from American and European written press and photographers from both continents. We were interviewed in turn, mainly our Commander Bill who showed our Hab around several times and explained the goals of our simulation and the experiments conducted. The media attention was also focused on the charming Andrea and Jan.

The MDRS 'open house' day for journalists and media representatives. *Credit* MDRS-5 crew

© Springer Nature Singapore Pte Ltd. 2018
V. Pletser, *On To Mars!*, https://doi.org/10.1007/978-981-10-7030-3_6

So much the better. In the meantime, I could get busy in the kitchen and serve warm drinks to everybody. I tried to cook a fresh loaf of bread to go with the remains of tuna with mayonnaise and the last bits of cheese, that were lying at the back of the fridge. I quickly turned the oven off when I noticed with Andrea a thick smoke that was coming from it. While choking our laughs to not disturb the on-going interviews, we discreetly unplugged the oven and put it in a cupboard. We'll see that later. I was reasonably left alone, until somebody remarked "Oh, but we have an astronaut candidate in our crew", and bang! that was it. I couldn't hide anymore behind my dish towel and in the fridge. I had thus to tell my story several times over and why it was important to prepare for the mission to Mars.

We had also foreseen to show how to don the EVA suits and the way we conducted an EVA expedition. With Jan, we kitted up and left the airlock to come face to face with a dozen cameras.

Kitting up for a demonstrative EVA with Jan in front of cameras. *Credit* MDRS-5 crew

In the front of the Hab (left) and climbing on the ATVs (right) in front of cameras. *Credit* MDRS-5 crew

Luckily, the weather turned back to sunny. We were asked to climb the cougar hill in front of the Hab, we were asked to pose for photos on top of the hill looking at the horizon, we were asked to walk and walk again several times for the televisions. And so on, until we mounted our ATVs and were followed from afar by the jeeps and vans of cameramen and photographers to go and pose further among the rocks. One of the photographers held us more than an hour looking into the sun while he was trying to take artistic pictures with colour effects. This little media circus lasted until five o'clock. And apparently, it had a rather impressive success. We were on the first pages of weekend supplements of several newspapers in the USA, England, Ireland, Spain, Germany, etc. Finally, the journalists and cameramen left satisfied with pictures and poses, and we could return to normal again. Quite friendly but rather tiring, and somehow invading after two weeks of isolation. We cleared up the Hab and, as the simulation was over, we treated ourselves to a walk outside, a real one, without EVA suits, only the six of us. We took the van parked behind a hill and we left on the *Lowell Highway*, in the direction of *Lithe canyon*, to wander searching for dinosaur traces. And we found them. Fossilized dinosaur bones, lying on the ground. Impressive and astonishing. It was there, on the ground, between rocks, not yet collected, and not yet in a museum. No: just there, in the wilderness.

The last outing out of simulation mode. Top: left, road to Henry's Hill; right: winding road; Bottom: left: Vladimir at *Dimitri's* corner; right: fossilized dinosaur bones in *Lithe* canyon. *Credit* MDRS-5 crew

We came home and started to prepare our last meal together. We wanted to do something special as it was our last evening together. We had three bottles of wine remaining and a few beers. As I was the DGO, I had the duty to prepare this last supper. With Andrea and Jan, we harvested our vegetables in the Hab and in the greenhouse. We did not find any bulbs, only stems and leaves. We cleaned them up, sorted them by kind, and arranged them in a salad as a starter.

The last supper: Top: Vladimir harvests the vegetables in the greenhouse (left); two of the eight dishes prepared (right). Bottom: a very proud DGO (left) prepares the dishes with the radishes, alfalfa sprouts, arugula salad and Chinese tatsoi cabbage grown in two locations, greenhouse and Hab. *Credit* MDRS-5 crew

The scientific experience was not yet over, as I had still to collect the impressions and opinions of everyone on the taste, the texture and the appearance of our homegrown production. In turn, each of us received a small portion of the eight salads and the four vegetables from the Hab and the greenhouse. It was served according to an Italian recipe: raw with a little bit of olive oil and salt and a glass of white wine. Simply delicious. A delight that we tasted with care and pleasure, while writing down our impressions. The rest of this five-course meal was another starter with fish (that is half a sardine per person, the last can), a dish of chicken sauté with rice, cheese (the last remaining zakouskis of the morning. Hey, we had to finish them all!) and for dessert, pears with chocolate. Not bad at all and well appreciated by the crew. Everything ended up in a good atmosphere, well after midnight. I finished clearing up everything and washing the dishes at around 2 a.m., satisfied about this great day.

The next morning, the last tidying up: throw everything in our bags and suitcases and loading the whole thing in the van. The last crew of this campaign of simulations would only arrive on Monday and we had to leave this Sunday morning to catch our respective planes. The hand-over would be done by Bill who was joined by his wife; both would stay around for another few days. We hit the road the five of us to go back to Salt Lake City. I fell asleep in the van and I woke up a few hours later in the middle of the mountains. We left the desert in the South to arrive in the mountain area in the North. What a difference! Snow on top of the mountains after all this heat! We dropped Andrea and David at the airport to catch their flights back at the end of the afternoon and we stayed with Nancy and Jan. The hotel where we stayed had an inside pool: what a delight, after so many days in the sand and the dust, to immerse oneself without having to count the seconds! We all went to bed early, after having said our farewells to Nancy who was taking an early plane the next morning. I watched CNN in my room and one of the top stories of the day was the French elections, with Chirac and Le Pen making it together to the second round. What a strange world we were coming back to!

Monday morning, we left directly for the airport with Jan, the last two Europeans from the crew to still be there. We had a hearty American breakfast and we parted our ways, swearing to see each other again. The return flight, via New York, was eventless. I could sleep a few hours in the plane bringing me back to Amsterdam. I arrived home Tuesday morning, took a shower, changed clothes, and went to work. Mountains of mail and e-mail were waiting for me at the office. Yes, life went on while I was on Mars in the desert, and I had to catch up two weeks of absence at work. And there I was, taken by the maelstrom of everyday life, of work and projects, each one more urgent than the other.

But what an experience again, unique and unforgettable!

On to Mars!

Chapter 7
What Did We Learn from This New Simulation?

Well, many things again!

First, that it was possible for a crew to live and work in harmony while in isolation during two weeks. Quite strangely, not one of us really felt isolated. Of course, we were cut off from the world and from what was happening there. But we were so busy with our various research tasks that we did not see the time pass by. And finally, these two weeks seemed to us so short, too short for what we had planned to do. I believe that this was the secret of the success of our rotation: we all had a program of scientific work that absorbed us so much that there was not a moment where we felt idle or lost. On the contrary, we would have liked to stay a few more days to finish up all our activities. In a certain sense, this simulation was more relaxed than the one of last year in the Arctic. We had the impression of having more time and, finally, we realized that these two weeks were not even enough. Another difference was the fact that we were freer in our schedules and in the way to conduct our investigations. In the Arctic, as our stay was limited to one week, we had to continuously hurry up to conduct all the scientific tasks and activities. This time, we could approach the whole thing in a more personal way, although we were working also together, as a team.

As already emphasized, one of the most important aspects of our EVA expeditions was the crucial use of the GPS system. It appeared also that it was not that easy to navigate and to find back the waypoints already indicated by previous crews. There is room for improvement for this aspect of field exploration. Several problems were outlined. First, the lack of reference to the used coordinate system: with respect to which origin and which earth model the archived coordinates were expressed? There are several tens of reference systems (and even several sub-systems in a same system) and these were not always indicated. This yielded much confusion and many differences of several hundred metres on the field, sufficiently to not be able to find back the target point. Then, to transcribe a point on a map is not sufficient to be able to find it back, as the way leading to it in straight line could not be feasible or could include impassable obstacles, necessitating a trial-and-error approach by several routes, which is not always possible when an

© Springer Nature Singapore Pte Ltd. 2018
V. Pletser, *On To Mars!*, https://doi.org/10.1007/978-981-10-7030-3_7

expedition duration is limited. Furthermore, it should be possible to confirm the point of interest by some information independent from its coordinates, e.g. an information on the visual aspect of the place to find, either as a text description or a digital picture, ensuring unambiguously that we have arrived at the desired place. The problem is therefore the archiving and the noting in a database and its relative size. Several attempts to clean up the database of visited points were done by the previous crew and our own. No really satisfying solution could be found, mainly due to the enormous amount of stored information, which shows the importance of the problem to establish a coherent program of exploration of a new area and, even more so, of a new planet. We need therefore geographers (or *areographers* ...) and experts in cartography during Mars missions, who could follow the first missions to put some order in all this.

Another practical problem that we had to face during this simulation was how to avoid contamination by the sand of electronic equipment and computers in the Hab as soon as the wind started to blow. And even more generally: how to avoid that the sand enters the Hab. Of course, our Hab was not pressurized and the sand could enter by the airlock doors and by the interstices of the windows and of the structure, as small as they were. But the problem would exist without any doubt for the Mars explorers, as the Mars dust would stick and stay on EVA suits and could also enter by airlock doors of a Martian station. We spent a lot of time at the beginning cleaning and dusting off the suits after each EVA outing and to clean the Hab the day before last. It is unconceivable to think that a Martian crew of four, five or six persons would spend several hours per day to continuously battle against dust and to clean up EVA suits and the various apparatus and instruments. This question is far from being solved and we think that it is not even properly asked in real terms. Nobody so far really realized the problem that the severe Martian dust storms of several days or weeks could cause, and the destructing and abrasive effect of dust carried by Martian winds. During our simulation, the damages to the Hab greenhouse and the continuous presence of sand inside the Hab were a constant example until we finally got used to it. Nevertheless, I noticed several days after I came back that I was still coughing up particles of this desert sand. Without consequences after a stay of several days, a longer stay of several months in such a dusty environment as Mars could pose severe health problems for the first Martionauts. Martian dust would be found everywhere and also in human organisms, lungs, eyes, ears, etc. An appropriate decontamination system must be envisioned to clean up EVA suits and airlocks by ultrasonic showers and to avoid dust coming into a Martian station, with sufficiently small filters. Here is some food for thought for engineers who work already on the design of the first Martian stations.

From a scientific standpoint, once more, the amount of collected results, information and data is considerable and we would need several months before being able to publish the results of biology experiments conducted around and in the Hab and of observations conducted in a desert so rich and varied from geological and paleontological viewpoints.

With respect to our experiment on plants, several trends showed up. However, these observations were not sufficient to be conclusive with respect to a decision regarding the design of a Martian experimental greenhouse. What was shown are the following points. The plants grown in the Hab were more attended and more cared for by the crew than those in the outside greenhouse. This result could have been guessed in advance, but nevertheless, it was well confirmed by this experiment during this simulation. The comments of the crew called upon the five human senses: taste, smell, touch, sight and audition. It showed clearly that some of us paid really attention to these little vegetables. Furthermore, it is interesting to see the comments made by the crew on these plants: some were simply interested ("Can we have sprouts for salad soon?", "Time to eat them!"); others more profound ("Very excited to see life in this rather "sterile" austere lifeless environment; only non-human life inside"), other recollecting past feelings ("Smell the ground to remind me the forest after rain at home"). From a psychological viewpoint, it is undeniable that to have a simple life form in the Hab, for which you have to care and feel responsible, was an important help to put up with the isolation. This is well known from some Russian cosmonauts on the Mir station who passed their free time taking care of a small garden with a few plants. It is therefore strongly advised to foresee to take green plants on board the first Mars missions. We also observed a cycle of two to three days in the intensity and frequency of comments and time passed to observe and care for the plants in the Hab, with an obvious important peak in the attention paid during the two days of the sprouting of the seeds. We still don't know if this a real natural effect, if the attention increases and then decreases every two or three days, or if it is due to external factors, like a busier schedule for other activities during days of less attention. Once again, a simple experiment during a simulation allowed us to notice behavioural and psychological factors, which are important aspects of isolation situations that Martian crews would endure during their travel and their stay of a total duration of several years.

Finally, the comments made by our crew during the tasting session of the different salads on the last day have also shown the importance for a crew to produce its own food, even if in this case, it was only a few shoots eaten as a salad as starter to a meal. It shows also that the sense of taste was not (yet) deteriorated. It is well known that the senses of smell and taste eventually blunt during experiments of long duration confinement and isolation, as reported by astronauts and cosmonauts having spent several months in orbit. For the record, certain salads were preferred to others, but without clear-cut results with respect to preference for the Hab or for the greenhouse grown vegetables. Nevertheless, it was the occasion for a celebration all together of the end and the success of our simulation.

Finally, it is really the most important conclusion from a psychological and simply human point of view: being able to show that after having the six of us spend two weeks locked in and isolated, there was no tension and bad feelings among us, we still got along well together. To the point where we could still enjoy a celebration meal together, even after two weeks. This makes us optimistic: one day, men and women would be able to do the same in a house on Mars.

Part III
The Desert Reload

Chapter 8
The Desert Reload—Before

Beside how to get to Mars and how to live on Mars (see the last part on this), there is also the question of how to conduct science experiments from the surface of Mars. Like all space agencies in the world, engineers and scientists at ESA start to think also about that aspect and more particularly how to conduct geology, biology and astrophysical experiments from the surface of the Moon and of Mars. This is not a simple question as we need to adapt and to optimize the existing scientific tools and instruments to enable good and productive science research from these extra-terrestrial planetary surfaces. Several aspects are changing with respect to an Earth-like environment: gravity-levels, atmosphere compositions, temperature ranges, humidity ranges, radiation exposure, electrostatic and abrasive dust environment, unknown bacterial environment (if existing), distance from Earth, human handling in EVA mode, etc. The list is quite long. This is one of the aspect that the project EuroGeoMars is trying to investigate at ESA, among others like assessing the development of scientific protocols and techniques in geology, astronomy and biology research in extra-terrestrial planetary conditions.

Dr. Bernard Foing, responsible for the EuroGeoMars project at ESTEC approached me in 2008 to explain that he wanted to organize a series of rotations at MDRS to investigate several human and scientific aspects of future manned missions on extra-terrestrial planetary surfaces for scientific exploration. He asked me if I could help him to set-up a new mission to MDRS and I immediately said "Yes!". The idea of going back to MDRS in the Utah Desert was very pleasing and we start immediately to work. The EuroGeoMars project was accepted by *The Mars Society* and would last for five weeks as follows:

- a technical preparation week (24–31 January 2009), for instrumentation deployment;
- the first rotation with the crew 76 (1–15 February 2009) for further deployment and the beginning of utilization;
- the second rotation with the crew 77 (15–28 February 2009) for further utilization and in depth analysis.

© Springer Nature Singapore Pte Ltd. 2018
V. Pletser, *On To Mars!*, https://doi.org/10.1007/978-981-10-7030-3_8

We continued to discuss with Bernard on how to organise this campaign from a practical point of view and who should be involved. We contacted several scientists from Europe and the US and invited them to join this effort. The response was quite overwhelming and we soon found ourselves with many requests for additional experiments. One aspect which was very important also is that we wanted to involve students in the practical preparation and running of the experiments. Today's students are tomorrow's researchers and it is a good thing to involve them as soon as possible in space research.

Two groups of experiments were eventually foreseen.

The first group included human crew related experiments, namely the evaluation of the different functions and interfaces of a planetary habitat, the crew time organization in this habitat, and the evaluation of man-machine interfaces of science and technical equipment.

The second group involved a series of field science experiments (geology, biology, astronomy and astrophysics) that could be conducted from an extra-terrestrial planetary surface and the necessary technology and networks needed to support them.

All these experiments were both interesting and exciting, and it was not obvious which one would be most appropriate in the MDRS environment. Due to my previous experience in the Arctic and in the MDRS, I was interested in the human aspects of crew operations. A young intern engineer with ESTEC, Ludivine Boche-Sauvan, would be the technical specialist responsible for conducting these human crew related experiments.

I would be responsible for the first crew 76 and Bernard would supervise the technical setting-up during the first week and the second rotation with crew 77. Further, I would supervise the work of Ludivine on all experiments related to the human operation aspects and, as she would be part of the second crew 77, the human crew experiments during the first rotation of the crew 76 would be carried out by me.

These crew aspects are important as, in manned space missions, the human factor is a dominant factor and may strongly influence efficiency and work results. To quantify such a difficult and uncontrollable aspect of space missions, it is necessary to reproduce as closely as possible the environmental and technical conditions in which astronauts would work: limited resources including power, communication bandwidth and manpower, limited social interactions in an isolated and cramped area, surrounded by dust, etc. The EuroGeoMars campaign in the MDRS was an ideal occasion to observe and measure these aspects.

During the mission preparation phase, we worked a lot with Ludivine to establish the approach that we would be following. We had consulted the similar work that was done earlier by other researchers during my previous simulations in

2001 and 2002, namely by Bill Clancey from the NASA Ames Centre and by the NASA-JSC Crew Support Office. This research formed the basis for the preparation of the questionnaires that would be used during these two simulations with crews 76 and 77.

Three axes were identified from this first research approach.

First, the evaluation of the different functions and interfaces of the MDRS Hab and its sub-systems (labs, crew quarters, rooms, computer and electronic area, common space, etc.) for the general crew usage and living (ergonomics, comfort, general set-up, etc.).

Second, the crew time organisation with respect to two aspects: how the crew members experience their daily organisation and how to optimize productive time. It was important to take into account how the crew members perceived their daily time organisation during the simulation, from wake-up until bedtime. The day was considered in periods of 1 h, in order to identify how to optimize the useful time periods and to diminish the wasted and slack time. The other aspect deemed important was to identify ways to optimize productive time (i.e. time spent on experiments and producing results, including EVAs) and to diminish the wasted but necessary time (repetitive tasks, maintenance, computer and instrument debugging and troubleshooting, etc.).

Third, the evaluation of the man-machine interfaces, i.e. how the crew members perceive how the various science and technical equipment (not only science instruments but also all other Hab subsystems, including EVA suits, radio, quads) are installed, used, stored, and manipulated in and around the Hab, and without and with EVA suit gloves. Furthermore, what would be the recommended changes to the man-machine interfaces to improve the daily operations of all Hab subsystems and instrumentation.

After several iterations, Ludivine finalised the questionnaires along these three axes which all crew members would be required to complete. The questionnaires were supplemented with time and location evaluation sheets to be filled in daily to support the reconstruction of the time spent at each task by all crew members.

In addition, the crew would participate in an on-going food study on the type of food imposed and crew impressions would be collected via questionnaires.

Regarding the second group of field science experiments (geology, biology, astronomy and astrophysics), several instruments were either brought from Europe or lent by US collaborators. Most were deployed and installed during the technical week.

The goal for the geologists in both rotations was to test different instruments and methods to characterise the ground and the subsurface layers in the extremely varied and rich environment around the MDRS Hab. Instruments like a Ground Penetrating Radar (GPR), a Raman Spectrometer, a Visible Near Infrared Spectrometer (VIS/NIR), a Magnetic Susceptibility Meter, and some drilling equipment would be lent by the NASA-Ames Centre. These instruments would be supplemented by an X-ray Diffractometer/X-ray Fluorescence Meter (XRD/XRF) lent by the *inXitu* company, and some additional sampling collection and curation, scientific and HDTV cameras for field and laboratory studies lent by the ESTEC ExoGeoLab, one of the many research laboratories at ESTEC. All these instruments were advanced, highly performant, relatively small and could be used in future space missions on or around Mars to detect underground water and to characterize the regolith and the different subsurface layers on Mars. Interestingly, the Ground Penetrating Radar would complement the investigations that I had conducted during my first simulation in the Arctic where a seismic method was used.

The primary goal of the biology investigations was the analysis of microbial communities living in the soil in the MDRS area. For this, an Adenosine Tri-Phosphate (ATP) Meter, also lent by the NASA-Ames Centre, and a Polymerase Chain Reaction (PCR) portable laboratory from the ESTEC ExoGeoLab were to be used.

For astronomy and astrophysics observations, we planned to use the radio-telescope of the MDRS Musk observatory.

For the engineering supporting projects, a rover, lent by the Carnegie Mellon University would be deployed along with a range of rover visualization test kits and camera systems with image data treatment kits.

So, as you can see, this is a very ambitious and extensive investigation programme in different fields. I have to confess that I am not an expert in all these fields and do not know how to use all of these instruments. That is why we have in both teams experts for each field. However, I have some understanding on how these instruments function and I will try to explain the use of some of them in the following pages.

Talking about experts, let me introduce those who would be part of the two crews and those who would support the science operations.

In the first rotation, with me as the Crew Commander, Euan Monaghan will be the Executive Officer (EXO). Euan, a brilliant Engineer, holds a Bachelor's degree in Physics with Space Science and Systems from the University of Kent, and a Master's in Astronautics and Space Engineering from the University of Cranfield, United Kingdom. Beside engineering and astrobiology, he also fences *épée*. He will be the Second in Command, taking care of the engineering and logistical tasks in the Hab, and of the astrophysics observations.

Jeffrey Hendrikse is also an Engineer but this time in applied physics from the University of Delft in The Netherlands. As an optical and nuclear engineer, Jeffrey worked on a variety of projects at ESTEC, in microgravity projects, on the first ESA's Autonomous Transfer Vehicle and on the Herschel observatory, a cryogenically cooled far infrared space telescope. He worked also on a radiation therapy project at the cyclotron in Groningen. Unfortunately, he would not stay he entire duration of the simulation as he would have to leave for Kourou, in French Guiana, to conduct the last test activities on the Herschel telescope before its launch on an Ariane-5 rocket. While at MDRS, he would conduct the technology and communication investigations.

Anouk Borst, a Bachelor student in geology at the Faculty of Earth and Life Sciences, at the Free University of Amsterdam, also joined the team. Anouk had already conducted several field investigation campaigns in The Netherlands, France, Germany, Belgium and Spain. As an intern at ESTEC, she analyzed the geology and mineralogy of the largest Lunar impact crater located on the far side of the Moon using datasets from Lunar missions Clementine and SMART-1.

A further team member from the Free University of Amsterdam in The Netherlands was Stefan Peters, a Master student in solid Earth geology. His thesis research on the coupling of impacts and volcanism on the Moon was performed at ESTEC and involved the mapping of lunar surface structures and comparison of mineralogy in different regions from multispectral data. As you could guess, Anouk and Stefan would be the two geologists in our rotation and would conduct all geophysics field investigations around the Hab.

Danielle Wills is South African and has lived in The United Kingdom. A Master student in astrophysics and in philosophy, she studied high redshift early universe objects. She worked also on various extracurricular projects in planetary science and astrobiology. As an intern at ESTEC, she worked on aspects of lunar astrobiology and the selection of landing sites of astrobiological interest for Mars landers and sample return missions. She further contributed to an astrobiology experiment flown on the International Space Station, and on testing the stability of halophilic archaea under simulated Martian conditions. And here you also guessed correctly, she would serve as the crew biologist conducting the field and lab investigations.

Pooja Mahapatra, of Indian nationality and a Space Master student at Luleå Technical University of Kiruna in Sweden, was experienced in designing and constructing CanSats and picosatellites alongside electronics and data processing for a project on fluid separation in microgravity. Also an intern at ESTEC in space engineering, she would replace Jeffrey Hendrikse after his departure and would be in charge of the engineering and technology investigations.

An interesting crew of talented young people with mixed and complementing backgrounds and expertise.

The second crew was equally extremely qualified and would be commanded by my friend Bernard Foing, an astrophysicist, space scientist, Professor at several universities, Chief Scientist in several space projects and in charge of the Research Division in the Science Directorate at ESA.

Ludivine Boche would be the EXO and the board engineer, in charge of the technical and logistical running of the Hab and also in charge of all the human crew related research.

Christoph Gross, a Ph.D. student at the Free University of Berlin in Germany, worked in the Planetary Evolution and Life research program and associated with the ESA Mars Express mission on the High Resolution Stereoscopic Camera (HRSC) project.

Lorenz Wendt, also a Ph.D. student at the Free University of Berlin, conducted several field campaigns in Argentina, Jordan, USA and Spain. Lorenz worked also in the HRSC Team for the Mars Express mission, integrating HRSC multispectral data and hyperspectral NIR data from two other Mars Express instruments and taking care of the geochemical modeling.

Pascale Ehrenfreund, from Austria, gained a Master degree in Molecular Biology from the University of Vienna, a Ph.D. in Astrophysics from the University Paris VII, and a habilitation in Astrochemistry from the University of Vienna. Currently involved in several international research projects in astrobiology, she lead the Astrobiology Laboratory at Leiden Institute of Chemistry in The Netherlands.

Cora Thiel, with a Ph.D. in Biology from the University of Bielefeld in Germany, worked on interdisciplinary biology projects including zero gravity research during ESA and DLR aircraft parabolic flight campaigns and on the International Space Station.

I hope that, by now, you have also guessed that for the second rotation, Christoph and Lorenz would be the crew geologists and Pascale and Cora would be the crew biologists.

Eight nationalities were represented in these two crews, as crew members came from Belgium, France, Germany, The Netherlands, United Kingdom, Austria, South Africa and India. This shows clearly that science has no national boundaries and that space research, and in particular Mars exploration, is by nature an international endeavor, at planet Earth level.

Prior to the simulation campaign, the EuroGeoMars group, including the two crews and the scientific and technical support teams, held several classrooms and simulated field training sessions at different locations in Europe for all the group members to get acquainted with each other, to train on the various instruments and equipment to be used during the simulation campaign and to rehearse investigation procedures and protocols. Several scientists and research engineers from universities in The Netherlands, France and Germany participated in these training sessions. In addition, Carol Stoker and Jhony Zavaleta (both from NASA Ames Research Center), and Philippe Sarrazin (a French man living in California, from the *inXitu* company) would give support at the MDRS during the technical preparation week, prior to the first simulation, to train the various crew members on some of the instruments to be used in the field.

These two rotations would not be an easy mission as the scientific programme was very ambitious and a lot of work would need to be done by all crew and support team members. After several months of preparation and training, we were coming close to the simulation dates of end of January 2009. We planned our journey slightly differently than that of seven years previously. Instead of driving to Hanksville from Salt Lake City in Utah, we would meet in Grand Junction, a relatively large city in Colorado and go from there to Hanksville and the MDRS.

However, we also learned bad news: Pooja's US visa had not come through and she would not be able to join us. Sad but true: Mars is still very far with regards to paper administration.

Crew 76 patch designed by Euan Monaghan. *Credit* MDRS-76 crew

Chapter 9
The Desert Reload—During

EuroGeoMars Crew 76 at MDRS: 28–29 January 2009

We arrived at Grand Junction (GJT), Colorado, in dispersed order last night (Wednesday 28 January). I arrived first from Salt Lake City (SLC), coming from Minneapolis and before that from Amsterdam. During the flight SLC to GJT, I saw that everything was covered by snow. It was eerily white all over. It was even worse up North, in Minneapolis, where it was about −20 °C with snow and ice covering the roads.

I arrived at GJT around 6 p.m. local time, which was already 2 a.m. body time (I had left home 8 a.m. on Wednesday the 28th). Anouk and Stefan arrived at 8 h 30 p.m. an hour later than foreseen and without Anouk's luggage. Euan arrived later after 9 p.m. We all packed up in the great white Japanese van that I rented and off we went driving to Hanksville. After a two-and-a-half-hour drive along the I-70 W from Colorado to Utah and a turn at exit 149 South on the UT-24, we arrived at Hanksville around midnight. Body time, it was already Thursday 29 January at 8 a.m. and we had been travelling for 24 h. After checking into the Whispering Sands motel, we crashed and managed a few hours of sleep.

I woke up early this morning but still in a fine form. It was a clear blue sky, with the sun shining on a very cold freezing morning. We repacked everything in the van and we cross the street to try to find a breakfast place. Unfortunately, the place, 'Blondie', was still closed and we went to the 'Hollow Mountain', a gas station next door, with the station shop inside the mountain, to grab a few sandwiches and cups

V. Pletser, *On To Mars!*, https://doi.org/10.1007/978-981-10-7030-3_9

of coffee and tea. We met there with Philippe Sarrazin, a Frenchman living in California, who would share his knowledge of the X-ray diffractometer instrument for geophysics. We went to find the Hab on a little trail road, aptly named 'cow dung road' leading to the MDSR Hab three miles further North. We arrived and found Bernard Foing, Jhony Zavaleta and Carol Stoker. Bernard showed us around the Hab, including the Observatory and the Greenhouse. I experienced the strange feeling of '*déjà vu*' as I could recognize many places where I have been seven years previously. Seeing the Hab again gave me also pleasant feelings. Pleasant because it reminded me of so many happy memories. But the Hab somehow had changed: loaded with so many cablings, electronic gadgets, papers, books, CD/DVDs, shelves, …

After a much appreciated quick sandwich lunch arranged by Bernard, we had two demos in the lab downstairs. The first one by Jhony Zavaleta who showed how to use the Raman spectrometer to analyse geology samples, using a small Infrared laser diode shining on the sample. Ingenious and potentially dangerous for the eyes if you are not careful with this little diode. And the other by Philippe Sarrazin who showed us how to use the XR diffractometer. Also quite ingenious and powerful, especially when I realise that you do not need a data connection between the instrument and the laptop, a wireless connection by wifi, it is all that it takes. Then it was time to leave for an exploratory geological walk with Carol Stoker as our geological guide. Anouk and Stefan, being both geology students, were of course in geologist's paradise as they were stopping every five meters to compare, admire, select, hammer down, pick up, etc. any pebble and other soil samples. An interesting walk that took us to the plain in front of *Skyline Rim*, and then South and back to the Hab. Unfortunately, Anouk twisted her ankle, but luckily, it was not too serious. On our way back, we met DG Lusko, the owner of the 'Hollow Mountain' gas station, and also the Sheriff of Wayne County where Hanksville is located. Quite a big guy but very quiet. He had come to deliver some packages to the Hab and he gave a lift to Anouk.

The Hab itself needed a good cleaning and some reorganisation. The dust comes in from the desert sand all around us and as it is extremely dry, even in winter, there is always dust everywhere in the Hab. So, regular vacuum cleaning sessions on the ground floor where the boots are kept are a must. Strict procedures are put into place to try to keep the Hab clean and dust free.

So, *voila* for this first day on Mars. Everything goes well but we are all busy with little tasks. I will sleep tonight in stateroom No 5 and move into Commander stateroom No 6 where Bernard is for the moment when he leaves on Saturday evening.

Have a good night on Earth and in Europe.
All the Martian best to all!
Vladimir

EuroGeoMars Crew 76 at MDRS: 30 January 2009

Yesterday Friday 30 January, I was DGO (Director of Galley Operations, i.e., the person who cooks, dish washes, take the garbage out, etc.) for the entire day and it was a lot of work. I spent all the morning cleaning the kitchen and washing dishes, throwing away old food and putting some order in the various cupboards. I prepared breakfast (simply slices of bread and cheese, ham and jam on the table with coffee and tea). And I tried to bake fresh bread using the Hab bread machine.

End of morning, as Anouk had still not received any news about her luggage (and it is obvious as we do not have any signal at and around the Hab for mobile phones) and as we needed to buy some more fresh food and meat (we are not yet in full simulation mode…), we took the car and drove to Hanksville with Anouk so that she could call about her luggage and buy any first necessities that she would need. At the market place, we collected some of the prepaid food for the study that we would participate in once in sim mode.

People in Hanksville are so nice and kind. There is no need to lock your car up and everything is left open. The lady at the counter allowed us to help ourselves to the food and did not check anything. It is true that in small places like these everybody knows everybody. Then we went to the 'Hollow Mountain', the gas station which has its shop installed in a cavern in the mountain. A nice and unusual place, ran by Don Lusko, who looks after the Hab and its power generator (more on this later). Then back to the Hab to start the lunch. Fresh tomatoes, sandwiches, coffees, teas and fresh fruits. Nothing extraordinary but well appreciated. Everybody had a good laugh at my poor excuse for a bread, which had not risen at all, although I had put yeast into the mix. Once again, it was time to clear the table and wash the dishes. And luckily for me, I could escape at 16 h while the others were working and being trained on the geological drilling machine. I rode the ATV up to the *Skyline Rim* plain, such fun to ride in the wind again on the same plain I had ridden seven years earlier. On the way home, I got lost as the old tracks had been washed away by rain, snow and erosion and I found myself riding the ATV into a ravine without what looked like any possibility of getting out. I did remember however that it would eventually lead to the main *Lowell Road* that would bring me back to the Hab.

Another moment of happiness was experienced when I saw an official map of the desert area around the Hab that had been drawn by crew 73 and I could recognize areas and features that we named with Jan Osburg way back in 2002. For example, *Dimitri's* corner (named after my son), *UFO landing site*, *Brussels sprout* (in the middle of a road), etc. It was good to see that efforts of pioneers were still remembered…

At 18 h 30, it was back to the DGO duties. I prepared a three-course meal: chicken soup; pork chops with rice, lentils and stir fried vegetables; and salad followed by a mixed fruit salad. Everything was on the table by 19 h 30 and I guess that everybody was impressed. I was as well, as cooking with only three small gas appliances is not an easy task. My second attempt at bread once more gained a laugh, as this time, although slightly better than my first attempt, it did not look much different from a geologist's rock sample.

Then it was back to washing dishes and cleaning until 22 h 00 where we then had time for some e-mails and report writing and a game of chess with Bernard. I will not tell you who won (but it was not me). News came in from Danielle and Jeffrey that they would be arriving at Hanksville at 23 h 30. Stefan and I went to pick them up. Finally, everybody was there and the simulation could soon start. Last e-mail check, anything urgent needed to be sent and it was 2 a.m. Pffffff, long day and time for bed!

Good night and Martian wishes to all!

Vladimir

EuroGeoMars Crew 76 at MDRS: 31 January 2009

Today Saturday 31 January, I woke up early at around 7 h 30, probably still the effect of the jetlag. I felt fine in the early morning but got so tired later on. After breakfast (Euan was the DGO today), we got some demonstrations on the rover and on the drilling machine. I worked to catch up on e-mails and reports and I have to confess that I had a quick 30-min nap at 11 h. But it was so good and much needed, this power-nap.

At lunch time, bad luck stroke in the form of a power outage, just as we were sitting down to eat. On checking the generator, it was not apparent what the problem was. We unsuccessfully tried to restart it but it shut down after a few seconds. At this point, we decided to eat lunch while it was still hot and to see to the generator afterward. Well to make a long story short, we still could not figure out what had gone wrong with this generator. It is actually the back-up auxiliary generator. The first two main generators had failed in the recent weeks. So, if this one fails for good, well, we would be stuck. And we thought that was that. But eventually, after a lot of trials, persuasion and kind words, the generator would restart, puffing and coughing but holding on. But then the inverter system of the Hab input would not start again and we thought that it was the end of the story.

With no mobile phone coverage at the Hab and only contacts with Mission Support by e-mail, and Anouk having still not heard about her suitcase (three days now), we drove back into Hanksville and asked Don Lusko whether he could help. He promised to pass by. Anouk called the airport again, where people were not sure whether the luggage had left Amsterdam or not (of course, it did as Anouk had cleared US customs in Chicago) and then they were not sure where it was in Chicago but they promised to speed it up (same answer as yesterday and the day before). The bag could have been anywhere between Earth and Mars for that matter.

On our return, Don Lusko arrived and fixed the generator: it had to be refuelled properly and the Hab batteries needed to be recharged for a couple of hours. So, we let the generator running and we cut all power in the Hab until the night came in, when the power was turned on and the Hab was on again and warm! The temperature in the Hab in the afternoon when facing the Sun can be quite comfortable, about 15 °C, but in the shadow or later afternoon, it can drop down to 5 °C or less. So here we were, all reports has been written and Commander hand-over was completed. I was now responsible for the ship and its crew for the next two weeks.

Have a good night on Earth and in Europe.
Martian best to all!

Vladimir

Log Book for January 31, 2009, Commander's Report

Today, we suffered a major power break down due to the Generator failure at lunch time (1 p.m. local). DG Lusko was alerted and passed by around 4 p.m. to restore the generator and to start to reload the Hab batteries which were depleted. Full power to the Hab was being restored gradually, step by step. This power breakdown affected the entire schedule of the day, but some science training had been completed.

A geological short test to measure geomagnetic susceptibility was performed. XRD samples could not be measured due to the power outage but some safety demonstrations were performed for the drill experiment. The Technical crew gave further training on the rover and lab instrumentation.

A quick biological reconnaissance of the Hab surroundings and a survey of the already acquired bio samples were carried out.

Our Hab Engineers checked the EVA suit subsystems (fans and radios) and put some of the items on charge, to prepare for rehearsal planned for tomorrow. Further introduction to the use of the ATVs and of the EVA suits were given to newcomers by the departing Technical crew.

Protocols and questionnaires for the various studies (food study, crew aspects, …) were discussed in the morning.

Handover was completed between Bernard Foing (Commander of the Technical crew) and Vladimir Pletser (Commander of crew 76).

Vladimir Pletser   Bernard Foing
Commander of crew 76   Commander of Technical Crew

Photo Diary

Sunset from the observatory
(Credit: MDRS-76 crew)

Others heading to the stars
(Credit: MDRS-76 crew)

Are there really more computers
than people? (…. Yes)
(Credit: MDRS-76 crew)

On to Mars!
(Credit: MDRS-76 crew)

| Drill familiarisation (some of us are more safety conscious than others) (Credit: MDRS-76 crew) | A nice example of cross-bedding (Credit: MDRS-76 crew) |

Log Book for February 1, 2009, Commander's Report

"Who needs to shred paper on Mars?"

Today Sunday was our first day in official simulation mode, although we agreed to still do some of the science operations in an out-of-sim mode. As all crew members had been present for some days and had participated in the technical set-up week, it was decided that crew members could start their Sunday morning according to personal schedule (which is a nice way to say that they could have a late morning if they wished so…). The first morning briefing was agreed for 9 h 30. The whole crew reviewed all the Health and Safety procedures in case of fire, emergency evacuation, medical problems, etc. The procedures for the different studies into which we were participating (food study and crew aspect study) were reviewed and first questionnaires were filled. We continued by cleaning up the Hab and by reclaiming some space on the upper deck. We found various office stationery items, including a paper shredding machine…. Who would need to take that to Mars?

The afternoon was spent in training in the field. A pedestrian EVA training was conducted close to the Hab for three crew members, followed by two reconnaissance expeditions. An ATV reconnaissance outing took place with a view to finding potentially interesting site for biology sampling, and a parallel reconnaissance was conducted by the geologists to explore an area representing the base of the Morrison formation made of river and channel deposits. Some petrified wood samples were collected for further analysis by XRD (X-Ray Diffractometer) and Raman techniques.

Next to maintaining and checking on the Hab subsystems, our two engineers also conducted some engineering investigations. Frame grabber and a streaming video server were installed on to a netbook laptop and were successfully tested with a handycam.

The evening would see the start of the cycle of internal seminars where crew members would present their respective research projects. Our first guest speaker, Dr. P. Sarrazin, will speak about X-Ray diffraction instruments for Mars exploration and terrestrial applications.

So, a busy day again on Mars.
Martian regards

Vladimir Pletser
Commander Crew 76

Photo Diary

The abseil (aka evacuation training)
(Credit: MDRS-76 crew)

Cross bedded conglomerates in Salt Wash Member
(Credit: MDRS-76 crew)

Factory Butte
(Credit: MDRS-76 crew)

We're not lost, honestly!
(Credit: MDRS-76 crew)

In the distance...
(Credit: MDRS-76 crew)

From Skyline Ridge
(Credit: MDRS-76 crew)

Log Book for February 2, 2009, Commander's Report

"One of these nights ..." (The Eagles)

Today Monday 2nd February was an excellent day full of discoveries. We got to learn about our direct neighbours. A scouting pedestrian expedition had brought back photographic evidence of traces of a cougar. Also, one of our crewmembers was at last reunited with her luggage that had gone missing in Chicago since last Wednesday.

Science wise, a first pedestrian outing by our geologists uncovers the mysteries of a large white sand lens North of the Hab. It was observed that small fractures were formed underneath the sand lens. Various clay and sand lens samples were taken for further lab analysis, which was conducted in the afternoon. XRD spectra were obtained and compared to databases. Water samples from the Hab water tank were taken and analysed to assess whether they were bio contaminated or not. Results showed that the level of contamination was acceptable for human consumption. A scouting pedestrian biology expedition was further conducted to characterise soil and vegetation at sites of geological interest to conduct correlated studies. A two-person ATV reconnaissance took place in the afternoon to scout for water-rich sites for future potential biological sampling and to prepare for future navigation experiments. In preparation of the EVA activities of tomorrow, all EVA suits were dusted and cleaned, helmets were cleaned, backpacks and radios were verified and batteries charged. The status of all Hab systems were declared good. All internal webcams were functioning and online. DG Lusko will change the external water tank tomorrow.

After yesterday's evening excellent seminar by Dr Sarrazin on XRD/XRF facilities on the future Mars Science Laboratory, we intended to have a first relaxing evening and to enjoy a DVD with the whole crew. A next seminar was planned for tomorrow evening. With the Eagles music in the background, we were also wondering whether one of these nights, we would enjoy such a peaceful evening on Mars.

Martian regards,

Vladimir Pletser
Commander Crew 76

Photo Diary

Touring with the ATVs
(Credit: MDRS-76 crew)

Through the microscope: plant
matter in quartz grains
(Credit: MDRS-76 crew)

Where are the geologists?
(Credit: MDRS-76 crew)

Where are the geologists?
(Credit: MDRS-76 crew)

Some botany in the field
(Credit: MDRS-76 crew)

This is why we close the door at
night...
(Credit: MDRS-76 crew)

Anouk: Warrior Geologist
(Credit: MDRS-76 crew)

Looking towards the Henry Mountains
(Credit: MDRS-76 crew)

Log Book for February 3, 2009, Commander's Report

"Come on Dad, I am just a geologist: I look at rocks and I speak science mumbo-jumbo ..."

Our third simulation day was a great and successful day. We had a three-person pedestrian EVA, two two-person ATV expeditions and various additional work done in the labs and in the Hab.

This morning EVAs took our two geologists and our biologist on a pedestrian outing from the Hab to the outskirts of *Skyline Rim* to look for geologic and biologic samples and to assess the differences and the difficulties of working under EVA suit and protocol constraints. Despite more difficult working conditions, several samples from higher stratigraphy were obtained and brought back by our first-time scientist EVA-ers to be analysed in the coming days. From a biological perspective, several distinct environments were scouted for varying density and nature of vegetation with a view to select a future EVA site for biological sampling.

The morning ATV expedition took the Commander and one engineer to reconnaissance several locations for future navigation experiments to be supported by automated mini-rovers. The afternoon ATV expedition was used to carry out a reconnaissance with one of the geologists of several sites with high geological impact to prepare for future ATV EVAs by our team geologists.

The samples of petrified wood brought back yesterday were processed with the XRD/XRF instrument and showed different chemical compositions for different colours of the samples. High peaks of Iron (Fe) and Manganese (Mn) were obtained for the black part, while those were absent on the white mineralized wood.

Significant for the engineering projects, a test of video streaming over Wi-Fi was successful. This is a primary component to obtain video footage from the rovers and eventually from cameras mounted on EVA suits. Post-reconnaissance analysis of information returned from yesterday's ATV expedition was performed with remote support engineers in order to find the best sites for preliminary tests of the Mars Navigation system.

All Hab systems were found to be nominal. The Musk Observatory computer was brought successfully on line. Webcams were troubleshooted and one of two is currently online. Analogue video feed from the Musk Observatory was tested and brought online.

Yesterday's DVD provided both insights and laughter for the entire crew. We watched two DVDs. The first one, "Postcard from the future" from past-MDRS crew member and film director Alan Chan, was excellent and gave us a new insight on how geologists can assess themselves (see above quote…). The second DVD ("Phantom from Space", from 1953) was so bad, that it was an excellent laugh as well.

This evening will involve more serious entertainment as our astrobiologist, Danielle Wills, will give a seminar on "Heating mechanisms in galaxy clusters".

So, with our heads in the stars and our feet on Mars, we wish you a good Martian evening on Earth.

Vladimir Pletser
Commander Crew 76

Photo Diary

Hab and Greenhouse under a
blanket of stars
(Credit: MDRS-76 crew)

Our three scientists setting foot on
Mars
(Credit: MDRS-76 crew)

A safe return to the Hab!
(Credit: MDRS-76 crew)

Gorgeous crystals of Gypsum
precipitated on Hab Bench
(Credit: MDRS-76 crew)

Looking for something
Commander?
(Credit: MDRS-76 crew)

Fresh cinnamon bread (and piece
of art) for breakfast
(Credit: MDRS-76 crew)

Euan up the loft, waiting for the next bucket during water brigade
(Credit: MDRS-76 crew)

Stunning view over Muddy Creek from Box Canyon
(Credit: MDRS-76 crew)

Log Book for February 4, 2009, Commander's Report

"Mmmh, this pumpkin pie is crusty ..."

For our fourth simulation day, we decided to take a step back and reflect on what had already been accomplished. The EVAs that had been planned yesterday were postponed until later and some more intensive work in the lab and on the computer took place.

In geology, the many samples and data were classified and results were put together and correlated with microscope pictures. In biology, sampling techniques were tested including a quart drill and a handheld shovel, to train the crew to collect samples at different depths. In addition, in situ measurements with an ATP (Adenosine Tri-Phosphate) Meter were conducted at several distances from the Hab up to 200 m. Results showed that none of the ATP levels indicated a dangerous contaminated environment. The levels fluctuated significantly from locations to locations within a factor of ten, but with a general trend to decreasing values with increasing distances from the Hab.

The engineering investigations involved the team engineers going to the *Skyline Rim* Repeater Point to assess the feasibility of installing Helium balloons for the Mars Navigation experiment. The assessment provided positive results. The replacement USB network adapter was received and will be mounted on the rover for further tests tomorrow. Investigations over the web streaming video were continuing with good results.

Hab systems were nominal, except for an unexplained power cut of 10 min around 5 p.m. Yesterday's seminar on "Heating mechanisms in galaxy clusters" was quite informative on why galaxy clusters are not as cool as the theory predicts it. Also, it has nothing to do with the quote of the day, as this quote refers to a statement made over lunch about a pie that stayed too long in the oven and created much laughter. Tonight's entertainment will be more mundane as we intend to watch a DVD; which one is not yet decided and will be one of the subjects of discussion over dinner.

From Mars to Earth, the thickness of a pumpkin pie crust…

Vladimir Pletser
Commander Crew 76

Photo Diary

Rock samples being investigated
under the microscope
(Credit: MDRS-76 crew)

Stefan working under Martian
colours
(Credit: MDRS-76 crew)

Not to scale?
(Credit: MDRS-76 crew)

Soil sampling
(Credit: MDRS-76 crew)

Factory Butte (please send all Ford advertising royalties to Euan Monaghan)
(Credit: MDRS-76 crew)

Hab at night from the edge of Skyline Bench
(Credit: MDRS-76 crew)

"Wow..."
(Credit: MDRS-76 crew)

Log Book for February 5, 2009, Commander's Report

"Thenardite? What is thenardite? ..."

Our fifth simulation day was busy inside and outside the Hab. Two EVAs took place in parallel this morning and a rover test went on successfully in the afternoon.

Our two geologists set the course of their ATVs toward *Candor Chasma* to collect more samples, primarily gypsum and salt efflorescence. These were analysed this afternoon and thenardite (a Sodium Sulphate salt, or $NaSO_4$ like Chemists like to call it, with four atoms of oxygen, one atom of sulphur and one atom of sodium), among others, was found by XRD in the salt precipitated in the canyon stream.

At the same time, our biologist and the Commander went on more or less in the same direction on a pedestrian EVA to collect underground samples from several depths (10 and 30 cm). This EVA was shortened due to a problem in the EVA suit system (malfunction of the air circulation system) without further consequences.

After the installation of a missing component yesterday evening, an outside functional test of the rover remotely controlled from inside the Hab showed that the rover functions were nominal and ready to support future EVAs. However, the rover computer was not processing enough CPU power for the required extra video streaming. Investigations to continue.

In addition to these field and laboratory investigations, crew aspect investigations were continued. Food study questionnaires and location and time evaluation sheets were completed every evening, to give insights on one hand on the diet that future astronauts would eventually follow and on the other hand, on how best to improve the Hab design and interior layout.

Hab systems were all declared nominal.

As the crew showed signs of fatigue during the lunch break after the work done so far, it was decided that the team deserved a short break, in the form of an afternoon nap. The choice of yesterday evening DVD fell on the classic "Outland" where Marshall Sean Connery defeated some villains on Io, one of Jupiter's moons. This evening, a seminar from Jeffrey Hendrikse on the science package on ESA Herschel spacecraft promises to be interesting and would complement the previous seminar.

Thenardite was found here on Mars. Do you have some on Earth as well?

Vladimir Pletser
Commander Crew 76

Log Book for February 5, 2009, EVA Report, Danielle Wills Reporting

EVA #3 Report
Time: 11.00–12.15
Crew Members: Vladimir Pletser, Danielle Wills
Site Location: Morrison formation, roughly 500 m from the Hab
Transit Mode: Pedestrian
Objectives: To test the soil sample collection procedure under EVA working conditions. This involves digging holes of 10 and 20 cm with the use of an auger and hand-held spade, extracting uncontaminated soil samples from these depths, securing them into sterile sample bags and documenting the location.

Lessons Learned: It is very difficult to avoid contamination of samples while wearing the large EVA gloves. It was decided that wearing a pair of nitrile gloves over these when collecting samples would alleviate the problem. The problem of the malfunctioning helmet (intermittent airflow) that the Commander experienced during the operation is being addressed by our crew engineer.

Report on EVA Suit and System Status

Out of the six EVA backpacks, none of them were deemed safe to use. In one EVA backpack, the cloth canvas was holding broken pieces of the plastic casing which provided a degree of containment. However, the air supply only functioned intermittently depending on the position of the operator wearing the backpack.

Out of the six backpacks, the lids of #2, 3, 5 and 6 were good, and the body of only #4 was good, which means that only one EVA backpack could be reassembled and used safely.

An attempt to repair temporarily the backpack bodies of #3 and 6, which would give another potential two usable backpacks, would be carried out.

The three others were damaged beyond repair possibilities at the Hab and needed to be replaced urgently.

Log Book for February 6, 2009, Commander's Report

"Few tasks are more like the torture of Sisyphus than housework, with its endless repetition: the clean becomes soiled, the soiled is made clean, over and over, day after day" So it was clean-up today …

And on the sixth simulation day, crew 76 decided to clean up the Hab. Armed with vacuum cleaners and courage, crew members positioned themselves strategically at various locations throughout the lower deck and living area and attacked simultaneously the dust, sand, filled bins, empty boxes and reclaimed successfully a lot of space on the lower deck floor. Yes, it was needed, and we are all surprised at the amount of space freed up.

Earlier in the morning, one of our crew members, Jeffrey Hendrikse, left to go back to Europe, as he is part of the launch supporting team for the ESA Herschel spacecraft due to be launched soon on an Ariane-5 from Kourou, in French Guiana. We wish him and all the Herschel team good luck.[1]

After the morning clean up, further work in the lab continued with Raman and XRD spectra analysis for most of the samples collected in the previous days. The plan to test and deploy the GPR (Ground Penetrating Radar), after having yesterday received the updated software, had been threatened by the strong wind blowing in the desert. However, it was bench tested inside the Hab lower deck to assess its functionality. Later on, the two geologists went on a scouting reconnaissance of an igneous field in search of volcanic ash remains. Extremophiles (blue-green algae) collected in snow patches yesterday have been studied with the microscope at high magnifications. Further, the Raman technique was investigated for application to the biological samples to obtain composition spectra.

As a result of yesterday's incident with an EVA backpack system, an investigation of all EVA suits and backpacks has revealed that some of them can no longer be used. Beside this, all the Hab systems are nominal.

Yesterday's evening seminar by J. Hendrikse on the science package aboard the ESA Herschel spacecraft was very informative. This new science tool will be available to astrophysicists for the next three to four years. This evening, the crew will have free time, which means that they may engage in any legal activity as long as it is confined to the Hab. My guess is that most of them will continue to work …

From a clean and working Hab, it is good bye from us all.

Vladimir Pletser
Commander, Crew 76

---

[1]ESA's Herschel Space Observatory was the largest infrared telescope ever launched. It was successfully launched on 14 May 2009 and remained active until April 2013, when its coolant was depleted.

Photo Diary

Who broke the door down? Oh,
it's our mini Ground Penetrating
Radar.
(Credit: MDRS-76 crew)

"Warn us if you want to come down."
(Credit: MDRS-76 crew)

Representing the laser massive...
(Credit: MDRS-76 crew)

Danielle's other talent
(Credit: MDRS-76 crew)

A sample returned from the field
(Credit: MDRS-76 crew)

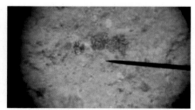

Examining minerals down the
microscope
(Credit: MDRS-76 crew)

Spot the geologist #2
(Credit: MDRS-76 crew)

Spot the geologist
(Credit: MDRS-76 crew)

Log Book for February 7, 2009, Commander's Report

"Good news! The Polymerase Chain Reaction (say simply Pee-See-Arrrh) equipment has arrived ..."

On our seventh day of simulation, two special events took place. First, we have a new crew member for two days in the person of Joshua Dasal, well known screen writer who produced 'Mars underground'. And second the PCR (Polymerase Chain Reaction) equipment has been delivered by the ever-present Don Lusko, after Fedex has made it travel back and forth in and around Utah.

A busy day with two parallel outings this morning, one for geology and the other for biology, and two other outings in the afternoon. Our two geologists went on to try out the GPR (Ground Penetrating Radar) system in the field this morning and afternoon on Hab ridge. Data were recorded and sent to the remote support team for further analysis, for the main purpose of instrument calibration. Further samples were analysed with the XRD and Raman instruments. Our biologist went with the rest of the crew to collect extremophiles like endoliths (living microscopic organisms living inside rocks or between mineral grains of rocks), and lichen samples around *Candor Chasma* for further microscope characterization and analysis in the lab. Further analyses with the XRD and Raman instruments were also performed. Another reconnaissance outing took our guest Josh Dasal and the Commander to explore *Lith Canyon* and the surroundings of the *Muddy Creek* River.

On the maintenance side, one of the ATV's broke down today: the gear foot selector lever had loosened on its axis due to it being worn out. Investigations on how to fix it are on-going with the help of Don Lusko. Troubleshooting and temporary fixing of the EVA suits and backpacks are also under way and could probably produce up to three EVA sets that could be acceptable for further temporary use. Beside this, all Hab systems are green to go.

Last night, the crew enjoyed a non-space related DVD, on request of our female colleagues. And we had the pleasure of watching Nicolas Cage being 'trapped in Paradise'. This evening, we will have a special lecture, as Josh Dasal has accepted to be our guest speaker and will present his movie 'Mars Underground'.

So, while we are looking forward to being taken underground on Mars, we wish you a peaceful evening on Earth.

Vladimir Pletser
Commander Crew 76

Photo Diary

The Moon rises above the Musk
Observatory
(Credit: MDRS-76 crew)

Astrobiology at work
(Credit: MDRS-76 crew)

An endolithic colony
(Credit: MDRS-76 crew)

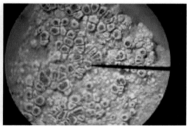

Lichen under the microscope
(Credit: MDRS-76 crew)

A geologist's playground
(Credit: MDRS-76 crew)

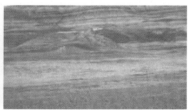

Some Martian wildlife makes an
appearance
(Credit: MDRS-76 crew)

Our guest for the next couple of days,
Josh Dasal, attacks the scenery
(Credit: MDRS-76 crew)

Log Book for February 8, 2009, Commander's Report

"If it comes out of the mouth of a Brit, it sounds more true ..." (J. Dasal, talking about 'Mars Underground')

This Sunday, our eighth day of simulation, was a relaxed day. The crew was authorized to wake up according to personal schedule, which means that we were all sitting for breakfast at 10 a.m., instead of 8. Although relaxed, we still did a lot of things.

A pedestrian EVA took place in the morning and a rock hunting expedition in the afternoon.

Our two geologists took our guest Josh Dasal for a nearly two-hour EVA tour of the Hab Ridge searching for geological samples. Rocks from the glistening seas and from dried out river beds were collected, including a strange specimen covered by an unknown yellow substance (more on this later on). XRD analyses are on-going.

In the afternoon, the whole crew went for rock hunting in *Lith Canyon*. Several large specimens of petrified wood were uncovered. Our biologist set-up all the PCR instrumentation in the morning and collected mud and soil samples in *Lith Canyon* during our afternoon expedition. These samples will be analysed by the PCR method overnight.

We saw the departure this evening of our guest Josh Dasal with sadness, as in just two days, he had managed to become an integral part of the crew.

On the maintenance side, our crew engineer succeeded to repair three EVA backpacks so we have now three functioning EVA suits and backpacks. We also experienced some strong gusts of wind this afternoon. The hatch protection on top of the Hab was blown off twice and luckily it was attached, so we could put it back in place and better secure it. Besides that, all Hab systems were nominal.

Last evening, the crew enjoyed a unique moment with Josh Dasal who showed us his film 'Mars Underground' in which he tells the story of the birth of *The Mars Society* with numerous footage of Robert Zubrin, Chris McKay, Penelope Boston and others, and some scenes had been shot at MDRS. We had a long and extremely interesting discussion with him on the making of the film (see the above quote ...). Josh also donated a signed DVD copy of his film to the Hab. So, let it be entered in the MDRS inventory. We believe that this film should be seen by all who arrive for the first time at MDRS.

After such an MDRS premiere last night, what shall we do this evening? Good night to all of you, Martians and Earthlings.

Vladimir Pletser
Commander Crew 76

Photo Diary

Extreme biology: Commander Vladimir holding Biologist Danielle
(Credit: MDRS-76 crew)

The life of an engineer... Euan and Vladimir fixing the roof hatch top
(Credit: MDRS-76 crew)

Goblin plus Euan
(Credit: MDRS-76 crew)

... and that was the last we ever saw of our Commander.
(Credit: MDRS-76 crew)

A petrified tree from the Jurassic era
(No, not the Commander walking above)
(Credit: MDRS-76 crew)

Log Book for February 9, 2009, Commander's Report

"There is no such a thing as a simple rock ..." (S. Peters, while the Commander was discarding a simple-looking piece of rock)

This Monday morning, our ninth day in simulation, we welcomed another journalist from Dutch television, but virtually this time. Our two geologists were interviewed by Skype early this morning for a Dutch TV popular science programme to explain why these Mars mission simulations are important to prepare future Mars missions.

After this, our now famous geologists and the crew biologist went for a sample collection outing in the lower blue hills plain. The geologists investigated the alluvial fan coming out of *Skyline Rim*, i.e., the regolith dragged away by flowing water and that deposited on smoother plain, in an attempt to reconstruct local climate and regional tectonic uplift. Some clay samples were collected as well. Other samples were further analysed with the XRD instrument in collaboration with a NASA scientist. The unknown yellow substance reported yesterday was analysed by XRD and Raman techniques but the obtained spectra do not match anything known, hinting that it may be organic instead of mineral. Analysis continues.

This morning, our biologist collected additional samples at several depths for analysis with the PCR instrument. This instrumentation was set-up, but still suffers from an inadequate power source interface and several alternate solutions are under investigation.

On the astronomy side, our resident astronomer, who is also our crew engineer, completed his certification for some aspects of the Musk observatory use, with the rest to be completed soon. Weather permitting, a penumbral lunar eclipse will be observed tonight. Furthermore, preparations are on-going to follow a Jovian radio storm this coming Wednesday. We hope to be able to use the MDRS radio-telescope set-up in front of the Hab.

From the engineering point of view, the rover computer is still undergoing testing to determine the most effective way to integrate it with the video streaming system.

Regarding the Hab systems, we suffered strong winds and rains today. The high winds caused the roof hatch to dislodge several times despite the improved securing system employed yesterday. We strapped it additionally hoping that it will not be blown away overnight. The Musk Observatory dome was investigated for a future refit crew. However, the securing mechanism for the dome in case of high winds could not be found. All other Hab systems are nominal.

After our magic evening of Saturday, the crew decided to have a break on Sunday evening and just enjoyed some quiet moments of discussion and e-mail. This evening, we will have a new seminar from one of our geologists, Stefan Peters (see above quote), who will rock us with some on-going debates in geology.

So, let Martian rocks roll us to Earth.

Vladimir Pletser
Commander Crew 76

Photo Diary

The offending hatch
(Credit: MDRS-76 crew)

It's life Jim...
(Credit: MDRS-76 crew)

The weather front approaches
(Credit: MDRS-76 crew)

... but not as we know it.
Reflexions of Stefan
(Credit: MDRS-76 crew)

Definitely an afternoon to be
working inside in the warm.
Anouk in front of a computer
(Credit: MDRS-76 crew)

The weather finally catches up
with us
(Credit: MDRS-76 crew)

Log Book for February 10, 2009, Commander's Report

"Let's do it again but this time in English! Take two ..." (F. Hubert, Belgian TV journalist during afternoon EVA)

Our tenth day of simulation was again a busy one. We had two new visitors, a TV crew who came from Belgium to record our activities for a popular science programme. They will stay with us for two days.

Our two geologists went on expedition this morning to collect samples in the Mount Rushmore area, hoping to find volcanic ash samples. According to all literature when comparing pictures and compositions, it appears they have. They managed also to bring some more samples that were further analysed with the same XRD and Raman techniques.

As a solution had been found to power the PCR supporting instrumentation, our crew biologist spent her entire day preparing and purifying samples and extracting DNA to be multiplied overnight by the PCR technique (see further).

On the astronomy side, the course for the observatory use was completed. The radio-telescope setups and controls were tested in view of the Jupiter radio storm foreseen for tomorrow Wednesday 11 February 2009.

The Commander and the EXO went on a two-person EVA to *Half Circle Ridge* canyon this afternoon, with the TV crew filming their activities (see above quote).

The ATV gear foot selector was examined again today with the help of DG Lusko and was found to be completely worn out and in need of replacement. Beside this, all Hab systems were go.

Yesterday's seminar by Stefan Peters was very entertaining and took us from the slopes of sedimentary basins where petroleum forms to the apparition of the first multicellular sea animals. Tonight, as the crew is still busy with reports and e-mails, it was decided that we would enjoy some chilling music while everybody is still working.

So, from a working and buzzing Mars, we wish you all a peaceful night.

Vladimir Pletser
Commander Crew 76

Photo Diary

A windy day on Mars
(Credit: MDRS-76 crew)

A mounted EVA
(Credit: MDRS-76 crew)

Exfoliation on volcanic tuff
(Credit: MDRS-76 crew)

Vladimir being interviewed while
washing up
(Credit: MDRS-76 crew)

Don't they need suits?!
(Credit: MDRS-76 crew)

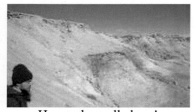

He was here all along!
(Credit: MDRS-76 crew)

Explicative Note

The PCR (for Polymerase Chain Reaction) technique is used in molecular biology
to amplify a single or a few copies of a piece of DNA (DeoxyriboNucleic Acid, a
self-replicating material present in quasi-all living organisms' chromosomes as

genetic information carrier) across several orders of magnitude, generating thousands to millions of copies of a particular DNA strand. So, it can be seen in a very simplistic way as a chemical "photocopying" process that allows to multiply the original DNA strands. It is an easy and cheap, but powerful, tool to amplify a segment of DNA that allows to improve the signal of identification of the different components. It was developed by Kari Mullis in 1983 and it earned him the Chemistry Nobel Prize in 1993.

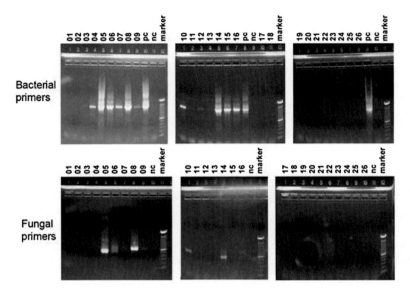

PCR experiment images obtained at MDRS by the following crew 77 (Credit: C.S. Thiel et al., 2011)

Log Book for February 11, 2009, Commander's Report

"We will experiment with sensors and software that will help us manage a generator and batteries that provide power to a habitat, while we are living and working inside" (Bill Clancey)

The eleventh day of our simulation was marked by a catastrophe and a miracle, or a death and a rebirth. Our brave auxiliary power generator that had been in use since more than two weeks died unexpectedly early this morning. The Commander and the EXO went to attend to it this early morning as soon as awoken and before breakfast (that was the catastrophe). We fought desperately to reanimate it but to no avail. We called upon DG Lusko's expertise and he came to declare that it suffered a final blow when one of the gaskets started to leak a massive amount of oil. Luckily, we still had some power stored in the Hab batteries but not for long. We were looking at running the Hab at minimum power, i.e., no heating, no warm water, etc. until the next day. DG Lusko came back in the afternoon with maybe more than an idea in mind. He managed to resuscitate an older generator, called Wendy, which is now working perfectly although it was also declared dead a few weeks ago. So, we are now back in full power swing mode and relieved that we would not have to face the freezing night with just our duffel sleeping bags and extra jackets.

Despite this power battle that took all day, we also had several other activities. A three person EVA took place this morning during which our two geologists and our crew biologist accompanied the Belgian TV crew to demonstrate the EVA capabilities for science field work. The two geologists ran a GPR (Ground Penetrating Radar) profile of an inverted river bed on Hab ridge to see whether the structure underneath the overlying Dakota sandstone could be visualized. Data have been sent to the remote support team for further detailed analysis. Our biologist collected more samples at different depths to run a complete PCR (Polymerase Chain Reaction) analysis over them. Later on tonight, we will observe the stormy radio emissions of Jupiter using the MDRS radio-telescope. As a result of the certification for the Musk Observatory use, the eye-pieces and camera systems have been received.

From a Hab point of view, the replacement of the auxiliary generator by the older Wendy system took most of the day and masked any other problem. Beside this major issue, all Hab systems were nominal.

After yesterday's chilled but work evening, we will have this evening another seminar from Anouk Borst on Moon geology. The two new guests, Francois Hubert and Tom Vantorre from Belgian TV, will stay with us tonight.

From a reinvigorated and powerful generator heated Mars, we wish you all a peaceful night.

Vladimir Pletser
Commander Crew 76

Photo Diary

The only picture that meant anything today...
(Credit: MDRS-76 crew)

Log Book for February 12, 2009, Commander's Report

"A lot of people like snow; I found it an unnecessary freezing of water...." (Carl Reiner)

During our twelfth day of simulation, the crew 76 once more conducted breaking edge science in the lab and in the field, despite snow falling in the afternoon.

Our two geologists updated their database of geological samples gathered so far and further Raman spectra were obtained in the lab. Our biologist went in the field like a bionic woman, equipped with a camera and a microphone mounted on her gloved hand and connected to a transmitting laptop installed in a backpack. She managed to collect new samples for further PCR (Polymerase Chain Reaction) analysis, while describing and allowing visualisation for a remote operator in the Hab of all the accomplished field operations. This technology experiment is intended to offer an enhanced real-time reporting capability of a crew member in the field to a Home base.

This afternoon, we received an incomprehensible phone call from Jupiter. No, it is not an alien joke. We set our radio telescopic ears up to 20.1 MHz to listen to the radio emission interaction between area A of Jupiter's magnetic field and its satellite Io. Not bad, eh! It gives another meaning to "Please, leave a message after the beep."

The Belgian TV crew filmed several everyday life scenes, from breakfast to working in the lab. Outside scenes around the Hab and at *Candor Chasma* were further shot before they departed.

All Hab systems were go, especially our new generator Wendy who behaved well, giving us full power.

Yesterday evening, we had a very interesting seminar from our geologist Anouk Borst on the geology of the Moon South Pole Aitken basin, and she nearly convinced us that we should go there to take some rock samples. This evening, it will be the Commander's turn to hold a seminar on spin-offs from space research or on weightlessness and microgravity.

We hope that all is going smoothly on Earth and from the snowy plains of Mars, we send you our good nights.

Vladimir Pletser
Commander Crew 76

Log Book for February 13, 2009, Commander's Report

"Happy Bread-day!" (Stefan Peters, while cooking his bread this morning)

During this thirteenth day of simulation, we started to take measure of the work already accomplished. Realizing that soon our stay here will be finished and we will have to leave the place to our colleagues of Crew 77 who will continue the EuroGeoMars programme, we started to put things in order. Procedures for using instruments, hints, and tips were put together, databases were completed, reports were written, and so on.

Samples are still being discovered and brought back for a last analysis, and new surprises appear again. Our two geologists and crew biologist went together to collect samples at *White Rock Canyon* and at *Mount Rushmore* below Hab ridge. Rocks with the same unknown bright yellow substance were found, and the amount of this type of (probably) salt yields a lot of questions. While geologists were scratching their heads, our biologists collected ten samples at different locations and at different depths of 10, 30 and 60 cm for further PCR (Polymerase Chain Reaction) analysis.

A two hour scouting and reconnaissance ATV outing took place this afternoon to explore the *North Pinto Hills* area to the east of the Hab. The Commander realized that there is more stuff out there to discover during the next simulation...

As we come to the end of our simulation, we started to fill out two general questionnaires on crew impressions and crew interfaces in order to obtain first inputs for future studies of planetary habitats. These questionnaires are quite long and intensive, which explains the unusual silence at this hour in the Hab. Everybody is concentrating on this important transfer of acquired experience during their two week stay at the Hab.

Regarding the Hab, there are no major changes to its status since yesterday; all looks, sounds and is OK.

Yesterday's evening seminar by the Commander on weightlessness and parabolic flights kept the crew awake with interest. Let's hope that the last seminar of our rotation this evening by our EXO/Engineer/Astronomer on "Mission to Titan" will score even better. From a slowly-winding down crew on Mars, we hope to see you all back on Earth soon.

Vladimir Pletser
Commander, Crew 76

Log Book for February 13, 2009, Commander's Report

"Little green men from Mars... And now glowing ..." (J. Pletser)

Crew 76 had some fun yesterday evening. It was Friday 13 after all. But the scare was real for a moment. It was all about these mysterious unknown yellow samples that have been baffling us for the last few days. We received an e-mail from Mission Support suggesting that we use a Geiger counter to verify whether these samples were radioactive, as there are Uranium ores in the area around the Hab. If it would be the case, these yellow samples could then be a Uranium salt known as "Yellowcake".

No need to panic on an empty stomach. We first finished our dinner and we started to think about what to do. Sample isolation, potential decontamination and cleaning procedures were devised and eventually applied. We wanted also to measure whether we really had a radioactive contamination in the lab, but we could not find a Geiger counter, just its empty casing. After having packed all sample material in double plastic bags and placed everything in a container outside the Hab, we found finally a Geiger counter that confirmed to us that the samples were slightly radioactive, less than 1 microSv/h (micro Sievert per hour, a unit of radiation measurement), which corresponds approximately to a dose that you would get on a high-altitude flight. So, no worries. Nevertheless, all crew members were instructed to take a shower last night and to scrub all over and particularly under the nails. So, no glowing Martians around...

Today being our last day of simulation, we spent a lot of time doing the last little things: taking pictures, clearing up, putting things in order, finishing yesterday's questionnaires, etc.

After lunch, we decided to go on a car expedition with our biologist insisting on going in EVA mode to collect her last samples. So we aimed for a three hour ride to *Skyline Rim* and all plateaus behind *Factory Butte*. Breath taking views of different mesas and underneath valleys. We found a spot were samples could be taken and we returned to the Hab to finish again all remaining reports and questionnaires.

Yesterday's evening seminar from our EXO/Engineer/Astronomer on "Mission to Titan" was skipped for the obvious reasons of decontamination and cleaning procedures. It is now postponed till another date. This evening, the crew will have a free evening allowing them to catch up with their tasks, chores, reports, questionnaires and so on.

Von Braun said once "There are two things standing between mankind and space: gravity and paperwork. I think we solved the problem of gravity". Hoping that paperwork will never be in the way of going to Mars, we wish you all Godspeed.

Vladimir Pletser
Commander Crew 76

Photo Diary

Our last day together as a crew
(Credit: MDRS-76 crew)

"Frank bedankt"
Stefan and Anouk thanking
one of their colleagues, Frank
(Credit: MDRS-76 crew)

On mars too, it is Valentine's day
(Credit: MDRS-76 crew)

The glorious Factory Butte
(Credit: MDRS-76 crew)

An unusually tilted Morrison formation
(Credit: MDRS-76 crew)

The aptly named Ghost Ranch (there was nothing there)
(Credit: MDRS-76 crew)

Chapter 10
The Desert Reload—After

Afterward

So, what happened afterwards? Well, as for my previous simulations, suddenly, it was time to leave and there was no more time to write down all that happened. But here is an accountant from my memory. The second EuroGeoMars crew arrived on the Sunday 15 February and each of us debriefed our respective counterpart of Crew 77. I had a long talk with Bernard, the Commander of Crew 77 to tell him all that happened, what we did and what we did not and the location of the slightly radioactive Uranium Yellowcake samples. By the way, I had received an email from Robert Zubrin in the night of Friday 13 to Saturday 14 February to confirm that these samples were absolutely without any danger. The four geologists and the three biologists were discussing happily respectively about rock samples and biology procedures. They had already started some work in the two different labs. Both crews together finished the Sunday afternoon with a long walk in the fresh air which was very welcome after two weeks of isolation. And then suddenly, it was time to pack up, load the van and leave. We drove straight to Grand Junction where we arrived in time to devour a huge steak and fall asleep in a motel before catching our respective flights back to Europe.

I left with strange and mixed feelings, as always. On one hand, sad to leave the Hab as again it was a tremendous and extraordinary experience to live together all these adventures for two weeks with this very talented crew, and on the other hand happy to come back to freedom and to 'normal' food. I was also somehow disappointed. The Hab had not changed much from what I remembered seven years earlier, which in a way was a good thing, but on the other side it did not improve. It was one of our major recommendations seven years earlier, to fix once and for all these little problems such as the power generator, the water pump system, the EVA suits, fans and battery packs, etc. This aspect was quite disappointing that after seven years since our first season of rotation, the Hab still needed so much attention and maintenance. My disappointment was also linked to the fact that I found it unfair to rely on the good will of volunteers and scientists who came to conduct experiments and investigations in an extreme environment, not to carry out

© Springer Nature Singapore Pte Ltd. 2018

V. Pletser, *On To Mars!*, https://doi.org/10.1007/978-981-10-7030-3_10

maintenance chores and DIY repairs. But, ok, if it was the price to pay (in addition to the 500 US Dollar that each participant had to pay…), so be it.

Summary of results

From an operational point of view, the work of our Crew 76 rotation can be summarized as follows. The field science experiments were started as soon as the corresponding instruments were assembled, tested and deployed. Several tens of samples were collected for geology and biology and were analysed in the lab at the Hab. It has to be noted however that the delay of one week in arrival of the PCR (Polymerase Chain Reaction) equipment due to US custom problems caused the postponement of the lab work on biology samples. Data were sent to remote science support teams in Europe and the USA for further evaluation and detailed analysis.

The geology investigations concerned mostly geochemical analyses of returned samples from the surrounding rock formations. For this, several advanced and miniaturized instruments specially developed for future space missions were used, including an integrated X-Ray Diffractometer/X-Ray Fluorescence Meter (Terra 158), a Raman Spectrometer (InPhotonics) and a VIS/NIR Spectrometer (OceanOptics). Approximately 50 samples have been analysed for chemical composition (XRF) and mineralogy content (XRD, Raman, VIS/NIR), varying from clay to sandstone and volcanic ash layers of the Jurassic Morrison formation, pure crystals such as gypsum and calcite, petrified wood, desert varnish and salt efflorescence. One of the field discoveries was the deposit of a bright yellow substance in a salt efflorescent soil. With extra caution, samples were taken and analysed with the available equipment. With the help of the remote science team and measurement with a Geiger counter, it was determined that these samples were Uranium salts known as Yellowcake. The sampling and analyses involved the set-up and maintenance of a detailed sample database with description of sample, context geology and test results. In situ magnetic susceptibility measurements were taken with a handheld magnetic susceptibility meter. Subsurface characterisation has been carried out with a miniaturized GPR (Ground Penetrating Radar), recently developed by NASA, to study the structure of a paleochannel (or "old channel", i.e. a remnant of a past river that has been filled or buried by younger sediments).

The primary goal of the biology investigations was the analysis of microbial communities living in the soil in the MDRS area. This investigation had a field aspect and a laboratory aspect: soil sampling was done in the field at depths of 10, 30 and 60 cm, in and out of EVA working conditions. DNA extraction and PCR (Polymerase Chain Reaction) analysis were then done in the laboratory. Samples collected for DNA extraction and PCR analysis included soils from various locations. The PCR analysis of these samples was delayed until the final necessary step, the agarose gel analysis. This was performed by the biologists of Crew 77, who continued the investigation for a further two weeks after Crew 76 had departed. The biology laboratory at the Hab only became fully operational at the beginning of the final week of rotation. Laboratory experimentation had been, due to the problems with US custom and unavailability of power transformers. Time only then allowed for the protocol to be successfully completed for the first sets of samples to be analysed. All investigations ran smoothly and efficiently in this minimalist laboratory, proving the concept that such a laboratory is sufficient for delicate science operations of this nature. Due to the necessity to analyse collected samples as soon as possible, the equipment delay also caused a delay in the field aspect of the investigation. The first week of the mission saw the biologist collaborating with the geologists, engaging in secondary biology investigations, and rehearsing the sample collection technique for the primary investigation in the field. The secondary investigations that were carried out included a biological contamination assessment of the Hab area, and brief analyses of snow algae, lichen and endolith samples in the laboratory. After running several trials with various digging and drilling devices, it was found that the most efficient way to extract soil samples from the depths required the combined use of a hand-held spade and an auger.

Habitat and operational maintenance took up the majority of the engineering time spent at the MDRS, especially during the period spent running constantly on the small backup generator. Maintaining this generator required daily manual refuelling as well as a minimum of two daily tests for oil leaks and other checks, reducing efficiency for the crew. The changeover to the more rugged Onan generator in the last few days of our rotation reduced this task substantially.

To mention some other engineering tasks completed: all six EVA backpacks were disassembled, inspected and their condition documented, with three deemed fit for purpose after repair. Towards the middle of our rotation the foot shifter on one of the ATVs failed; an investigation discovered that the unit was worn-out and the replacement took place after we left the MDRS. A significant amount of time was lost resolving computer issues, with most time spent on internet issues relating to intermittent connection and poor upload bandwidth.

In addition to keeping the Hab in power, water and sanitation, two main engineering projects were conducted while at the MDRS, both of them related to the streaming of remote audio/video. The first of these was the assessment of a tele-operated rover system, on loan from Carnegie Mellon University, to allow upgrade to high-resolution streaming video, and the second an in-field audio/video device for remote assistance and documentation. Excellent groundwork for the second EuroGeoMars crew was laid in both of these areas. A third project, the Mars Navigation System, was delayed due to an absence of necessary materials, with only some preliminary reconnaissance conducted.

From a field operational point of view, only eight EVAs were conducted for geology, biology, technology and reconnaissance purposes. The reason for this low number of EVAs was twofold: the poor state of the EVA suits and backpacks and the fact that only two ATVs were available from the beginning of the rotation, one having failed after one week.

From a human perspective, Crew 76 started with six crew members. However, Jeffrey Hendrikse, who left after one week, was not replaced by Pooja Mahapatra, who, due to visa problems, was not allowed to travel to join Crew 76. Three journalists also occupied the Hab: Josh Dasal (USA, screenwriter) on 7 February, and Francois Hubert and Tom Vantorre (TV, Belgium) on 11 February. The more than two weeks spent in the Hab in semi-confinement and semi-isolation were positive in developing bonds between crew members and one could observe mutual support and collaborative spirit increasing with time. No psychological problems were observed and a strong team spirit had developed.

Social activities were conducted as a group. All meals were prepared by one person in turn, this person being in charge for the entire day of all kitchen chores. All meals were taken together and were the occasion for planning, briefing and debriefing outings and crew activities. Other team activities were conducted in the evenings, alternating seminars presented by each crew member in turn, and watching DVDs or listening to music.

Lessons learnt and recommendations

After the Crew 76 rotation, we had a chance to talk it over again and we came up with a series of lessons learnt and recommendations that you will find here below, some are simply obvious.

The lessons learnt (in italics hereafter) during this first rotation can be divided in five parts: Habitat resources, Habitat layout and equipment, EVA procedures and equipment, crew day-by-day life organisation, and crew human aspects.

One of the major difficulties of the Hab's functioning was that of the amount of time required by the crew to repair or fix certain subsystems and to alleviate to the poorly functioning of some other subsystems. As crew time is the most unique and important mission resource, especially for qualified researchers conducting field science, it is difficult to justify crew members spending time performing daily chores due to failure of Hab subsystems.

A planetary habitat or a research laboratory should be designed such as to minimize the use of crew time to perform Hab maintenance. Crew members should be called to perform maintenance or repair only in exceptional circumstances and not as part of a baseline daily routine.

(1) Habitat resources

The most important mission resources at the Habitat are power, water and communication capabilities.

For power, the MDRS Hab depends on Hab batteries (located under the Hab) and on external diesel generators (located in the external engineering area). Due to previous failures of the two main generators, an auxiliary diesel generator was used during the first ten days of the Crew 76 rotation, needing daily diesel refuelling and oil monitoring. This auxiliary diesel generator failed several times, causing several power cuts, but could always be restarted by the crew engineer. After ten days in the rotation, the auxiliary generator failed definitely overnight and could not be restarted, leaving the crew relying only on Hab batteries in a limited power use mode (i.e. without heating, hot water and additional electric appliances) in the cold with temperatures close to 0 °C. The local Hanksville support technician managed to restart one of the two main generators at the end of the day of the failure occurrence.

A planetary habitat or a research laboratory cannot function without continuous power. A reliable power source, either generator, or power plant, or connected to main electrical power network, should be considered an essential part of the habitat.

A large water tank is located in the external engineering area and connected to the Hab water piping system. However, the external piping system got blocked due to freezing. An auxiliary water tank (approx. 500 litres) was temporarily located near the Hab and was refilled regularly by the local Hanksville support technician. The water from this tank could not be pumped in the Hab pipe system due to a broken pump that could not be replaced. This situation required the crew to carry, by hand every other day up to ten large buckets of water from the external tank to the inside 60 l tank located in the loft. This nearly daily operation was called the 'water bucket brigade' and took 30 min to one hour. Using crew time to refill an inside buffer tank is not the most optimum way of using qualified manpower and should be avoided, except in exceptional circumstances.

The "water bucket brigade" at work. *Credit* MDRS-76 crew

Heating the water for consumption, dish washing and showers was done using the Hab electrical power.

A planetary habitat or a research laboratory needs a continuous supply of warm and cold water, and should be designed such as to provide, within specified daily limits, running hot and cold water without requiring the entire crew to perform refilling tasks.

For normal operations during rotations, the crew relied on e-mail communications with an external virtual Mission Support Centre and Remote Science Support teams. The Mission Support was virtual in the sense that members of the Mission Support teams were not located in one specific place but scattered all over North America (USA and Canada) and were in contact by e-mails only for a limited period (one hour) per day. No other external communication means were possible (there was no mobile phone signal available at MDRS).

Connections to the Internet (and thus e-mails) depended on several LAN or Wi-Fi servers in the Hab which were connected to a satellite antenna dish. Two technical problems regularly dogged the mission. The first was the limited contracted available bandwidth (1.5 Mb/s for download and 365 kb/s for upload, with a maximum amount of 300 MB transfer per day). This limited bandwidth precluded the transmission of large files which impacted drastically on the exchange of scientific files and information between the field crew and the remote science support teams, affecting the quality of science operations. The second was the regular failure of the transmitting antenna or the unavailability of a signal, which caused all communications to be stopped. This obviously is unacceptable in an environment where leading edge field science must be conducted.

Therefore, a Hab or research laboratory needs a reliable internet communication system, with a sufficiently large bandwidth to support science operations and data exchange with remote ground support teams.

(2) Habitat layout and equipment

The MDRS Hab was deployed in 2001 with limited private funds. Although the overall Hab concept is adequate, some of its subsystems and internal layout need to be regularly maintained or replaced. The lack of modularity in certain subsystems was apparent in the overlaying of several 'temporary fix' solutions on top of one another throughout the years (e.g. see further about LAN and Wi-Fi server routing).

This accumulation of temporary solution layers meant that the overall functioning of the Hab and crew comfort (and sometime safety) could be called into question and could certainly be improved and optimized (comfort here should be understood as providing the bare minimum for the crew to rest adequately at night and a normal temperature inside the Hab).

– Heating systems

The desert environment can be cold in the winter periods, with temperature close to 0 °C during days and freezing overnight. A new heating system had been installed, but besides generating heat, it also generated quite a lot of noise. This noise was bearable during days but hampered sleep at night. During power cuts, the heating system was off leading rapidly to an uncomfortable cold environment. During the day, even with the heating system on in the lower deck, the temperature was barely above 0 °C, requiring crew members to work with thick layers of clothing and sometimes gloves even for operations necessitating fine handling and manipulations.

To provide the necessary minimum crew comfort with respect to noise and ambient temperature, the design of a Hab or research laboratory needs to include silent and efficient separate heating systems for lower and upper decks.

– Upper deck

• LAN and Wi-Fi server connections

The LAN and Wi-Fi server routing (see photos) demonstrated an evident lack of modularity.

The design of a Hab or research laboratory needs to incorporate sufficient modularity to allow for upgrades over time of the main subsystems such as internet connection.

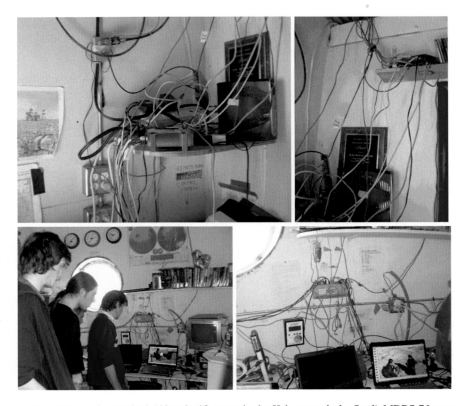

Internet cable routing to the LAN and wifi server in the Hab upper deck. *Credit* MDRS-76 crew

Compare the above situation in 2009 with the one during a previous 2002 simulation. *Credit* MDRS-5 crew

- Computer table

During a simulation, there were more computers than crew members. This situation was unavoidable as every aspect of Hab system monitoring, experiment preparation and data analysis required separate computers, which lead to an overcrowding of the computer table. Crew members were often found working on their laptops at other locations (in their personal stateroom, on the main central table, in the lab, …). Part of this problem could be traced to the fact that some Hab systems have evolved with time and their monitoring was done by introducing additional monitoring circuits and computers on top of previously existing systems.

To avoid overcrowding with computer and electronic systems, the design of a Hab or research laboratory needs to incorporate sufficient modularity to allow for upgrades over time of main subsystems for monitoring of the Hab systems and experiments.

Several views of the computer table with overlays of computerized systems. *Credit* MDRS-76 crew

- Kitchen facilities

The Hab kitchen corner included cupboards, a double sink, a water heating system, a fridge, a cooking gas stove without a cooker hood, and a microwave oven. An air conditioning system was installed above the stove. Dish washing was done by hand by crew members in turn using tap running water. Biodegradable soap was used to respect ecological constraints of the sceptic tank.

View of the kitchen sink and working table before dish washing. A crewmember washes dishes in the kitchen sink. *Credit* MDRS-76 crew

To save crew time and minimize water consumption, dish washing by hand should be avoided and an automatic dishwashing system (i.e. a simple dishwasher) should be included among kitchen facilities.

- Stateroom facilities

The six staterooms are quite small due to limited space in the Hab upper deck, but sufficient to provide space for sleeping. As temperatures dropped down drastically at night, winter sleeping bags were used, placed on a small wooden bench. Some bare minimal facilities were present in each stateroom: a limited lighting, a (very) small desk and some hooks.

View of a stateroom. *Credit* MDRS-76 crew

To improve crew comfort, stateroom design should be improved within the limited space to include an additional light source, small cupboards and drawers to store personal belongings, and a mattress.

- Loft and roof hatch

Above the six staterooms, a loft space provides some room for additional stowage. A roof hatch initially provided additional lighting through a circular transparent plastic window. This window was destroyed by strong winds over the years and was replaced by a more solid wooden case. This wooden case was blown away several times by strong stormy winds and was strapped down to the roof structure and to the loft floor. In one instance, the recovery of this wooden case during a storm necessitated a crew member held by another crew member to climb through the hatch.

A view of the roof hatch strapped to the roof structure and to the loft floor. *Credit* MDRS-76 crew

To improve crew safety and to avoid potential dangerous situation, the roof hatch should be designed to be secure in strong winds.

- Paper archives and office items

Over time, seven years of MDRS operations, many books, manuals, maps, reports and other paper archives have accumulated in the upper deck. Most of these were placed on shelves above the circular computer table, but sometimes on the computer table itself. In addition, a lot of stationary items, electronic gadgets, and other office material littered the circular computer table.

Books and manuals on shelf above the computer table and stationary office material and other papers next to computer on the computer table. The kitchen fridge is partially visible on the left. *Credit* MDRS-76 crew

In an effort to free some space, books, manuals maps and reports were placed in boxes and stowed in the loft area. Although, the importance of keeping archives and important paper documents (Hab system manuals, maps of the area, …) was recognized, the accumulation over the years made it nearly impossible to distinguish between important documents and old reports. In addition, it contributes to potential fire hazards.

To save crew time and to provide sufficient working space in a limited working area, important paper archives (Hab manuals, maps, …) should be scanned and stored in digital form on a centralized Hab database.

– Lower deck

• Geology and biology labs

The geology and biology lab area was spacious enough to allow simultaneous working in both disciplines. However, packing boxes and storage sometime filled the floor of the lab meaning a difficult access to certain parts of the lab and, more important, rendering evacuation exits sometime difficult to reach. In addition, the lab floor, made of metallic plates bolted to the underneath wooden structure, did not allow for a functional working environment for two reasons: first, the accumulation of outside dust and sand on the floor transported by shoes and winds through openings, and second thermal insulation could not be achieved with the metallic floor structure. Furthermore, the accumulation of samples and of various unused lab material over the years and of packing boxes on the floor did not ease the work of field scientists in the lab.

To improve the efficiency of lab work and the crew comfort and safety, the lab design should be such as to provide: efficient means to avoid accumulation of outside dust and sand in the lab; enough storage room for boxes and required lab equipment, an optimal temperature and regular inventories of all lab equipment and eventual discarding of unused lab equipment and samples.

Demonstration of the X-ray Diffractometer (left) and of the Raman Spectrometer (right) by US collaborators for geology on a table in the geology lab. *Credit* MDRS-76 crew

Views of the biology section of the lab. *Credit* MDRS-76 crew

Views of the biology (left) and geology (right) labs. *Credit* MDRS-76 crew

- EVA preparation room and Airlocks

The EVA preparation room was used to stow all EVA equipment (6 sets of simulated suits, helmets and backpacks; see further for comments on equipment). The
room itself could accommodate one crew member kitting up with two helpers, but
no more. If several crew members need to kit up, they had to do it in turns and then
wait in the lab area, which contributes to introduce more dust and sand into the lab.
Both airlocks for EVA (at front) and engineering (at back) have a high step on the
inside door, while both outside doors were needing to be fixed and could not be
closed properly.

*To improve the cleanliness of the Hab lower deck, the design of the airlocks,
outside doors and EVA preparation room should be such as to provide efficient
means to avoid accumulation of outside dust and sand in the lower deck.
Furthermore, the design of the EVA preparation room must be large enough to
allow more than one crew member to kit up completely.*

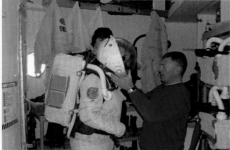

View of the EVA preparation room (left) and crew members kitting up (right). *Credit* MDRS-76 crew

Crew members waiting in the lab area (left) after kitting up (right). *Credit* MDRS-76 crew

- Bathroom and toilet

A small bathroom with a sink, a shower and some stowage cupboards are located next to a separate single toilet. To spare water, hot showers (connected to the kitchen water heating system) were taken in turns by crew members, every third day (i.e. two showers per day were allowed). All washing was done with biodegradable soap and tooth brushing was done with baking soda to respect ecological constraints of the sceptic tank. Running water was not available in the bathroom sink and in the toilet. Cold water buckets were used for washing when not showering, and in the toilet for cleaning. See above recommendations for water.

(3) EVA procedures and equipment

To conduct outside field work in simulation mode, crew members wore simulated EVA suits, including a helmet connected to two air pipes in which outside air was blown by a pair of fan located in the backpack. For long EVAs, a water pouch was placed in the backpack, connected to an outlet in the helmet. Communications were made through small portable radios attached to the helmet, either in VOX mode (voice activated, not practical) or in PTT mode (Push To Talk, activated by a button installed on the helmet supporting collar ring). Thick gloves and boots complete the ensemble. To activate keys on instrument keyboard, a finger extension (nail, pencil, …) was optionally tapped on a glove finger. When EVAs were made with ATVs, a rear-view mirror was optionally mounted on the forearm of the suit. These simulated EVA suits were representative enough to simulate the lack of mobility, the lack of finger precision and difficulties of manual dexterity, the limited field of vision, etc. encountered during real EVAs and were used to train crew members to the different ergonomics of working with EVA suits and gloves. Depending on experience, it took up to 30 min to 1 h to one crew member to kit up completely, including preparation of helmets (soaping vizier surfaces to avoid fogging), radio checks, glove finger extension and rear-view mirror.

ATVs were used to extend the range of exploration and reconnaissance EVAs. For obvious safety reasons, EVAs always included at least two crew members and crew members were always in radio contact between them and were monitored by a Capcom in the Hab, who logged all events occurring during the EVA.

– EVA suits and backpacks

EVA suits at MDRS were either a one or a two-piece suit. It was found during the first rotation, that out of the six suits, three were unusable as they were torn and worn out. Similarly, gloves were torn and worn out. In five of the six backpacks, five casings were found broken and had been repaired or fixed many times with tape. Fans, powered by rechargeable batteries, worked intermittently as there was not enough pressure difference between inside and outside the backpack casing. In one instance, during a pedestrian EVA, a crew member had to interrupt the EVA as his air supply was stopped by non-functioning fans.

To improve the field work efficiency, the crew member safety and the EVA representativeness, the design of simulated EVA suits and backpacks must be such as to have enough modularity for maintenance and regular replacement.

– ATVs

During the first rotation, only two ATVs were present at the Hab, instead of the minimum of three ATVs. One of the ATVs failed after about a week (foot gear selector jammed) and was temporary fixed until eventually it failed completely and the ATV became unusable. The other ATV got a slow puncture in the front tyre, necessitating the need to adjust the tyre pressure regularly.

Having only one ATV hampered the crew to go on long distance EVAs as crew members were not allowed to drive ATVs alone. A minimum of two crew members were needed to go on far away EVAs using ATVs.

To improve the field work efficiency and the EVA representativeness and to allow long distance EVAs, a minimum of three ATVs should be available in working conditions.

A crew member on one of the two ATVs, while they were still both functioning. *Credit* MDRS-76 crew

(4) Crew day-by-day life organisation

The day-by-day life of a crew of six persons living together in a limited volume required some organisation and discipline that needed to be managed firmly but with enough flexibility to accommodate individual needs and work constraints and to allow for unforeseeable major events occurring at the Hab (generator failure, water bucket brigade, …). This aspect of crew time organisation was the subject of the crew aspect studies. Only certain relevant aspects are highlighted here. Everyday chores were shared in turn. The crew also regularly spent time together during meals, morning briefings and evening debriefings, questionnaire filling, evening events (either seminars or watching together DVDs or sharing informal discussions). Preparation of reports was carried out individually, usually between 18 and 20 h.

– Everyday chores

The most time-consuming chore was related to kitchen activities and food preparation. To minimize the time impact on the daily working time, it was decided that one person per day will be responsible in turn for all kitchen activities and food preparation. This person was called the DGO (Director of Galley Operation) and was responsible for laying the table before all meals (breakfast, lunch and dinner), preparing and cooking the meals, washing dishes after meals, cleaning up table and kitchen, and taking care of kitchen garbage bins (that were left in special containers outside to be brought back to Hanksville by the local support). This entailed up to

2–3 h per day (excluding meal time) of chore work in addition to normal research work duties. The day DGO had the prerogative of choosing the music that would be played on the Hab sound system (if no other crew member would veto the choice of music). Despite the time spent, it was found that this method had the least overall impact on the working time over a period of six days. Other chores included the refill of the inside water tank (performed by the entire crew every other day with a duration of up to 30 min to 1 h) and the refill of water buckets (duration 25–30 min once or twice every day) to be used for the washing (when not showering) and for toilet cleaning.

To minimize the time spent by crew members to perform the necessary chores, the Hab design should include means to alleviate the amount of domestic work, e.g. a dishwasher should be available in the kitchen area, automatic pumping of water from outside to inside tanks, etc. (see above recommendations).

A crew member on DGO chore in the kitchen. *Credit* MDRS-76 crew

- Time spent together by the crew
- Meals and briefings/debriefings

The crew spent all their meals together, from early morning (breakfast) until late evening (post dinner dessert). These occasions were taken not only to socialize but also to hold the briefings (during breakfast) and debriefings (during dinner).

A meal shared by the entire crew. From left to right: Euan, Stefan, Vladimir, Danielle and Anouk. *Credit* MDRS-76 crew

- Daily questionnaires

Various daily questionnaires were filled individually but at the same time, i.e. all crew members were invited after the evening dinner and before dessert to bring their laptop to the main table and spend the 15–20 min to complete the questionnaires together. This approach alleviated the boredom to repetitively complete questionnaires day after day with a touch of humour and fun.

- Evening activities

Every other day, every crew member was asked to prepare and give a seminar for 45 min to 1 h over a subject of choice. This seminar was given at the main table using a personal laptop or the whiteboard, while other crew members shared the dessert and an evening cup of tea. DVDs were watched together at the beginning of the rotation, and with time, were replaced by free-time evenings were crew members could engage in either discussions, or personal work, or experiment reporting, or an activity of choice.

This approach was eventually favoured by the entire crew as the seminars gave the occasion to dwell upon subjects unknown to the majority and to appreciate the various expertise of other crew members, leading to increased understanding and mutual esteem.

Stefan giving a seminar on the whiteboard. *Credit* MDRS-76 crew

– Report preparation

During a rotation, *The Mars Society* organisation committee required several reports to be prepared either daily or every other day. The mandatory daily reports are: the Commander check-in report (reporting on the crew overall health and performance and on the main Hab system status), the Commander report (on the various day activities), the Engineer check-in report (on the detailed technical status of the Hab system status). Other reports, although optional, are strongly recommended and have to be submitted every other day: Science reports (on performed experiments and preliminary results), EVA reports (after each EVA, on duration, range, activity, results, interpretations, …), journalist reports (on more anecdotic aspects of Hab life or expedition), and a selection of up to 6 photos of the day's activities.

These reports and photos were sent by e-mail between 19 h and 20 h to the virtual Mission Support who reviewed them and placed them on *The Mars Society* website for the public to enjoy. *The Mars Society* used these reports to attract the public interest, attention and support. However, the daily preparation of these various reports was extremely time consuming (about 1–2 h per day) and their need is questionable. Besides the obvious communication aspect and the visibility that *The Mars Society* gets from having its website visited daily by the public, from an operational point of view, the only report that is needed on a daily basis is the Engineer check-in report on all detailed Hab system status. All other reports are 'nice to have', but not necessary and should be left to the discretion of the crew members, especially the scientists who should not be pushed to write something every day about their on-going experiments and preliminary results. Furthermore, the writing of reports on the computer and uploading them on the website were time consuming, considering also the limited bandwidth and the intermittent unavailability of the satellite connection.

To save crew time, the number of daily reports should be reduced and kept to a necessary minimum. A single engineer check-in report should be sufficient. Furthermore, instead of typing a report on a computer, audio files could be recorded and sent to Mission Support and leaving Mission Support to type and edit the report as necessary for upload on the website.

Preparing reports usually ends up in a rush before (and sometime after) dinner. *Credit* MDRS-76 crew

(5) Crew human aspects

This aspect was part of the study and only some relevant issues are highlighted here.

From a psychological point of view, no detailed account could be given as no psychologist was part of the crew and there was no external monitoring by psychologists. However, it can be stated that crew members did not have any interpersonal problems. Some issues surfaced and were taken care of by the entire crew. Group dynamics was positive and was enforced by adhering to collective group activities (meals, evening activities, …), by sharing chore responsibilities (DGO, water bucket brigade, …) and by facing together unforeseen events (generator failures, …). However, one aspect did threaten the group dynamics. A romantic relationship between two crew members had been established prior to starting the rotation at MDRS and it appeared that the ups and downs of the relationship did affect the other crew members. This kind of situation should be treated with a lot of caution and preferably should be avoided, either by not selecting the two crew members together (i.e. having them selected separately in different crews) or by engaging frankly with both crew members, and reminding them of their duties and priorities during the simulation. Another potentially threatening situation occurred when a crew member announced that he was considering leaving because the scientific equipment needed for the experiment was late (stuck in custom) and that the crew member had other research priorities that were awaiting him at his home institution. The situation was solved when another crew member intervened and persuaded him to stay on.

A last aspect involved privacy issues and the use of webcams. The Hab had several webcams, both in the lower deck and in the upper deck, that broadcast permanently on the web. It was decided that the webcam on the upper deck would be switched off to respect crew privacy during off-time,

Some results of the human crew aspect study

The results of the crew aspect study could be summarized as follows.

For the time and location evaluation, thirteen typical activities were followed: sleep, bathroom/toilet, common meal, common activities (briefing/debriefing/talks/evening activities), work in stateroom, work inside the Hab, work outside the Hab, EVA preparation and outings, chores inside, maintenance inside, maintenance outside, driving van and shopping at village, and non-EVA outings. Regarding the common activities, breakfast starting times and durations were approximately constant, except for both Sundays 1 and 8 February, at approximately 9 h and lasted approximately 45 min. Morning briefings were held during breakfasts and rarely lasted longer. Depending on operation and Hab failure constraints, lunch started between 13 and 14 h and duration was between 45 min and one hour. Dinner started also between 19 h 30 and approximately 20 h 30 and lasted for approximately one hour. Evening group activities started usually straight after dinner and ended between 22 h and midnight. The total time of daily common briefing/debriefing/discussion, including daily briefing during breakfasts, amounted to 17 h 55 min for the entire two-week rotation.

Other time parameters of significance are the duration of sleep and the time spent carrying out chores and maintenance. Averaging the sleep duration for the six crew members over the two weeks (only one week for one crew member) yielded an average of approximately 8 h 30 min which could be attributed to a hypothetical average crew member. One of the scientists slept the longest period (12 h) after a late arrival at MDRS to recover from jet lag. The Commander slept the least (6 h) on the first day upon arrival at MDRS. Some crew members enjoyed from time to time afternoon naps to recover from jet lag; durations of these naps are included in the data.

Regarding the chore duration, on some days, the crew member responsible for the daily chores was helped by another crew member. On the first day, the Commander spent the most time (4 h 45) tidying up the Hab after changeover from the technical crew. Remarkably, a certain periodicity can be seen during the first week, while the chore duration tended to decrease over the second week, showing again a possible more pragmatic approach to the chores, mainly in the kitchen. The average time spent per day on chores was approximately 3 h without help and 3 h 15 m with external help, which can be considered as a lot of unproductive time wasted on daily chores. However, spending time preparing meals for special occasions (birthday, celebrations, etc.), although not contributing directly to productive scientific work, does promote enhanced bonding among crew members, which can be seen as having a positive psychological effect.

Regarding maintenance duration, the EXO spent most of his time maintaining the Hab systems as it was his main responsibility, spending a total of more than 36 h. Other crew members spent between 40 min and 1 h 40 min on chores. The hypothetical average crew member whose average maintenance time would be the average of the six crew members, i.e. 1 h 23 min, which can also be considered as a lot of unproductive time.

Summarizing the average unproductive time, one could say that a hypothetical average crew member would sleep an average of 8 h 26 min, eat breakfast during an average of 44 min, lunch for 48 min and dinner for 57 min, spend an average of 3 h 08 min doing chores and 1 h 23 min doing maintenance, and spend an average of 1 h 35 min on evening common activities, which sums up to 17 h 01 min, leaving only approximately 7 h for work. Even if this average estimation is very crude, it shows that the remaining average time for work is significantly low and that a lot of time is wasted on unproductive tasks, chores and maintenance mainly.

Working with, and listening to, scientists develop an understanding of how they work. Scientists appear to adapt their methodology to the obtained results. The schedules can therefore not be predictable. In future missions, adaptation of laboratory schedules is a challenge to be faced.

Regarding crew interface and space occupation in the Hab, several remarks were made by all crew members. It would be too long to report them here. But it triggered an analysis of the working area in the lower deck and the traffic within it. The lack of space felt by the scientists in the laboratory is mainly due to the multi-use of this room. It is a biological laboratory, a geological laboratory, and also the central room in the lower deck, leading to large traffic between the experiments and sample analysis.

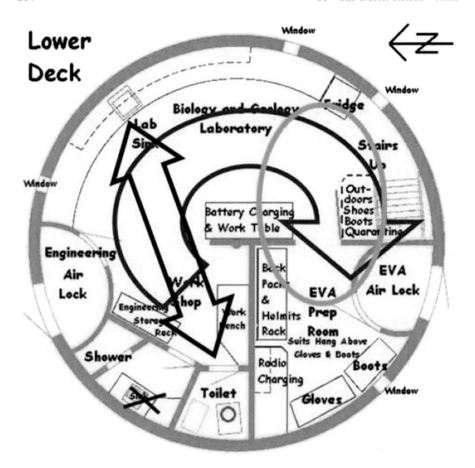

MDRS lower deck areas analysis. *Credit* Ludivine Boche-Sauvan

Traffic between the stairs and the bathroom, between the toilets and the lab sink (as the bathroom sink was not usable, the lab sink was mostly used leading to waste of time between different uses), and for the engineers, from the stairs to the engineering air lock for the daily checking and from anywhere to the workshop for fixing things (for example, from the EVA preparation room to the workshop to fix the EVA suits). This lack of space led to an uncomfortable working place, requiring the need to move items or to store them under the tables, which in turn did not allow crew members to sit properly and increased the occupation in the traffic area. This lack of space was not productive and it presented a safety issue. For example, shoes left downstairs at the bottom of the ladder on the floor to avoid dust upstairs, or not used during EVAs, were potentially dangerous for the scientists who could stumble over them on to sharp instruments or tools. Moreover, the organisation was important among the scientists, as they did not have the same needs: geologists needed darkness to use the Raman spectrometer and needed to crush samples for analysis with the X-Ray diffractometer/X-Ray fluorescence meter, while biologists needed a good light for soil kit analysis and a minimum of dust. In order to share the use of the single room laboratory, both groups of scientists had to work during the night alternately.

A lack of space in a planetary base is usual, as its overall design, mass and dimensions are limited by the size and capability of the launcher. However, improving the layout would allow a better use of its internal space. A first idea is to optimize the laboratory space with a dedicated place for each activity. Plastic sheet partitions could suffice to separate the geology and biology areas to allow geologists to crush their samples and biologists to work on bio samples in a less dusty environment. Other examples and recommendations were also given and can be found in several publications made after the simulations (see references at the end).

Some (preliminary) conclusions

It was shown from analysis of crew time and location evaluation sheets that the main unproductive day periods are in order of duration sleep, daily chores, collective evening activities, maintenance, and meals. As sleep, collective evening activities and meal time cannot be compressed in order to provide enough rest and relaxing periods to the crew, meal time were optimized by holding informally the general morning day briefings during breakfasts and the general evening day debriefings during dinners. More specific briefing and debriefing were held during the day as called by operation constraints.

This leaves that the main 'time eaters' are caused by daily chores and maintenance, to which should be added off-nominal maintenance due to Hab system failures. On average these consumed about 3 h per day for chores and about 1h30 m per day for maintenance. These could be drastically reduced by improving the internal layout and the design of the Hab and its subsystems. For example, kitchen chores could be reduced simply by using a dishwasher and water handling chores could be completely avoided by an appropriate system of pumps. Nominal maintenance could also be reduced by adapting the internal layout and design to specific working needs of crew members. Off-nominal maintenance could also be reduced to a minimum by implementing some of the recommendations.

Crew comfort is an issue that cannot be ignored. A Mars mission will last for typically three years. So, comfort must be provided to the crew to work and live in the habitat for this period. Again, comfort here is not meant to be the equivalent of a five-star hotel but as to provide the bare minimum to allow expedition crew to rest and relax properly in between working periods.

Finally, despite all the logistic difficulties and technical problems encountered, and the amount of time wasted on unproductive tasks, a lot of science results were obtained which reflected on the dedication and the expertise of all crew members. This bodes well for the future of Mars human exploration missions.

Part IV
Mars Tomorrow

Chapter 11
Mars Tomorrow

So, what has happened on Mars in the last 15 years? Where do we stand now? Are we ready to go? This chapter will try to bring some answers to these questions.

First, our knowledge on Mars has improved tremendously since the early 2000s with several iconic automatic missions: Mars Global Surveyor (1997), Mars Odyssey (2001), Mars Exploration Rovers (2003), Mars Reconnaissance Orbiter (2005), Phoenix (2007), Mars Science Laboratory (2011), Mars Atmosphere and Volatile EvolutioN (MAVEN, 2013), from NASA; Mars Express (ESA, 2003) and more recently, ExoMars Trace Gas Mission Orbiter (2016) from ESA; and Mars Orbiter Mission (MOM also called Mangalyaan) from the Indian Space Research Organisation (ISRO), the first successful interplanetary mission form Asia. Several landers and rovers explored also Mars surface, NASA Spirit and Opportunity rovers (on Mars since 2004), Curiosity (on Mars since 2012). In the near future, in 2020, a Chinese rover and another European ExoMars rover will explore the planet's surface.

<u>What did we learn that we did not know before?</u>
Well, the presence of water has been confirmed several times, and in different locations. On the polar caps, of course as water ice which is mixed with CO_2 ice, i.e. the ice formed by the precipitation of the atmospheric carbon dioxide under very low temperature in the higher latitudes.

© Springer Nature Singapore Pte Ltd. 2018
V. Pletser, *On To Mars!*, https://doi.org/10.1007/978-981-10-7030-3_11

In 2004, the OMEGA (Visible and Infrared Mineralogical Mapping Spectrometer) instrument on ESA Mars Express confirmed the presence of water as ice in the South Pole, mixed with CO_2 ice, with these three images taken at different wavelengths corresponding (from left to right) to water ice, CO_2 ice and visible. *Credit* ESA

Then as ice under the surface of the planet, which was discovered by the radar on Mars Express. It was also discovered as traces attached to other compounds in Mars' soil. Furthermore, the study of the geologic features on the surface of Mars has shown that liquid water must have flowed in the past. The HRSC (High Resolution Stereoscopic Camera) on board Mars Express has produced several tens of thousands of high resolution photos, some turned into videos. Nearly all of them show surface features most likely formed by past flowing liquid water.

Picture of *Reull Vallis* taken by HRSC on ESA Mars Express on 15 January 2004 from an altitude of 273 km (size 100 km, resolution 12 m/pixel) showing a sinuous ground feature, most likely an ancient riverbed. *Credit* ESA/DLR/FU Berlin

Furthermore, in situ examination of some mineral forms by NASA's Spirit, Opportunity and Curiosity rovers have undoubtedly established that these minerals formed a long time ago in presence of liquid water.

So, these are strong indications obtained from the recent European Mars Express mission and NASA Mars missions (Mars rovers, Mars Science Laboratory) that water could have existed in the past in a liquid form at the surface and could have possibly harboured some form of life. Finding evidence of a past or present life form is the Holy Grail of space scientists and is one of the main justifications for present automatic missions and future manned exploration.

The composition of Mars' atmosphere was already known to be 96% of carbon dioxide, with a bit less than 2% of argon and less than 2% of nitrogen and with traces of oxygen and water vapour. Surprisingly, methane (known also under its chemical name CH_4, i.e. one atom of carbon (C) to which four atoms of hydrogen (H) are attached) was also discovered in the atmosphere by the Mars Express mission and confirmed by other observations. Methane was observed to appear in several discrete (or point like) locations on the surface of the planet. Methane has a residence time of about 300 to 600 years in Mars' atmosphere which makes it a relatively recent occurrence.

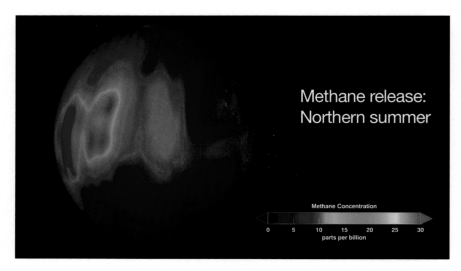

Concentrations of methane discovered on Mars. *Credit* NASA

The *Nili Fossae* graben system is an area of great geological interest. A high methane concentration was discovered above this area. Part of a large, 55 km-diameter impact crater with a central pit can be seen on the top right. This image was acquired by HRSC on 16 October 2014 (resolution about 18 m/pixel; North is to the right, east is up). *Credit* ESA/DLR/FU Berlin

Although the quantity is relatively large, its origin is unknown. And its presence is a mystery. Nobody really understands where this methane is coming from. Atmospheric methane is usually the sign of two kinds of activity: it can be produced by volcanic and geologic activities or by biogenic or organic activities. On Earth, methane exists also as a trace gas in our atmosphere. On one hand, increased values correspond to volcanic activities; during a volcano eruption, lava, rocks, fumes, and gases, including methane, are emitted. On the other hand, the more or less constant background of atmospheric methane on earth is produced by the biosphere, and mainly by ruminants. But, don't misunderstand me, I do not want to say that there are cows on Mars, no!

Like Earth, Mars has also volcanoes. Mars has even the tallest one in the whole Solar System, *Olympus Mons* at 20 km altitude, about two and half times higher than the tallest mountain on Earth (*Mount Everest* at 8848 metres). However, they are all extinct since long and there are no active volcanic or geologic activities on Mars.

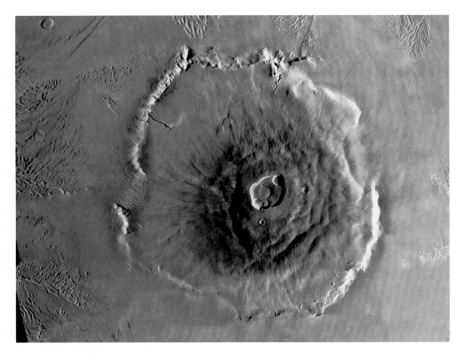

Olympus Mons. *Credit* ESA/DLR/FU Berlin

Does this mean that there is biological activity on Mars? Not necessarily. In fact, scientists are extremely cautious and are examining all possibilities. Among these, two could provide some explanations, but even these are not satisfactory enough. The first one involves a slow geochemical reaction, called serpentine reaction. This reaction occurs between olivine rocks (a type of iron containing ore that is relatively common on Mars), water and carbon dioxide to produce another type of rock known as serpentine and methane. The second is clathrate hydrate, a compound which consists of a lattice of host molecules (e.g. water ice) that traps a guest molecule (e.g. a methane molecule). Clathrate hydrates also exist on Earth and are believed to form most of the permafrost in Siberia. It is feared that due to climate changing heating, water ice in the permafrost could melt and release tons of CH_4 in the atmosphere which would accelerate its heating. Now a similar process of trapping methane molecules by water ice molecules could also exist on Mars. However, the various numerical models for the production of clathrate hydrates, serpentinite rocks and biological activities do not explain the amount of methane measured in the Mars atmosphere. The large amounts of methane in Mars atmosphere remains a mystery.

Another surprise was the discovery of several holes at the surface of Mars. NASA's Mars Odyssey orbiter observed up to seven holes near the extinct volcano of *Arsia Mons*. These are most likely a collapsed roof of a lava tube. Lava tubes are created when volcanos erupt. Lava flows as liquefied rock in fusion and this more or less cylindrical flowing lava mass cools down from the outside to the inside. When the lava production ceases, the centre part of the more or less cylindrical tube continues to flow while the outer layers are already solidified. So, you end up with natural caves having the shape of an approximately cylindrical tunnel. With time passing by and erosion, probably water and wind at the beginning and then later, only wind, the upper part, the roof of the lava tube collapses, exposing the empty space underneath.

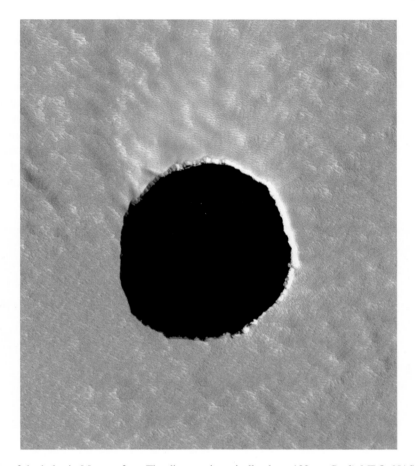

One of the holes in Mars surface. The diameter is typically about 100 m. *Credit* **MRO, NASA**

These holes are potentially extremely interesting. First, if there is life (still conditional as no life has ever been confirmed), it is possible that it could have migrated under the surface, at a certain depth under the Mars surface regolith to escape harmful radiation at the surface. These lava tubes offer a natural protection and it is only natural to think that microbial life, if existing, could possibly be found at the bottom of these holes. Second, if human crews were to be sent to Mars, these natural caves would provide a perfect protected environment from the surface radiation. As these caves may be too difficult to access by rovers, studying them in situ by human astronauts is an interesting possibility and they could become the location of the future first human colonies.

It would be somehow ironical though when you think of it that it took millennia for the prehistoric humans to come out of terrestrial caves, to evolve to a technological level allowing mankind to travel to another planet, to eventually seek refuge in caves on another planet. Well, history may sometime go in cycle after all.

Talking about mysteries, let us put an end to a well believed false one: the face on Mars. In 1976, the Viking-1 and -2 orbiters returned many photos of the surface, including the one below on the left of an area called *Cydonia Mensa*. At the first look, it resembles strangely, although vaguely, a human face. From this, some science-fiction authors, conspiracy theorists, and other crack-pots came up with many stories. One recounted that Mars was populated by starving people and that this face sculpture was a cry of help to Earth's mankind to come and rescue them. Another story told of how it was a leftover sculpture of a visiting alien party. Since these first photos of Viking probes, several missions, including Mars Express, have revisited Mars and the *Cydonia Mensa* area and with more sophisticated, higher resolution cameras (such as the HRSC) and other instruments. They have revealed that it is in fact a hill whose shape and surface are sculpted by wind erosion.

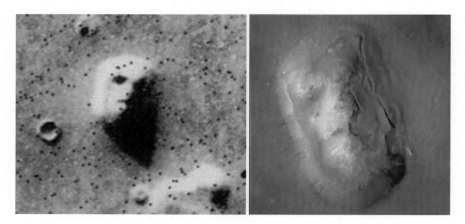

The "Face on Mars" has proven to be a hill structure in the *Cydonia Mensa* area on the surface of Mars; compare the left photo taken by NASA's Viking-1 orbiter in 1976 (*credit* NASA) and the right one taken by the HRSC of ESA's Mars Express 40 years later, in 2006 (right, *credit* ESA).

There is so much still to say about what has been discovered on Mars by these NASA, ESA and ISRO missions. The interested reader is invited to consult the books or websites given in the references list. Much more is still to come as other missions are in the planning. The second part of the ESA's ExoMars mission is planned to be launched in 2020, with a rover having a drill capability that will allow soil samples from down to two metres below the surface to be collected. The Martian regolith provides enough protection against radiations at two meters and the hope is still to discover some microbial form of life that could survive in the harsh Martian environment.

Artist view of the ExoMars rover to be launched in 2020. *Credit* ESA—AOES Medialab

China announced at the end of 2016 that it will also launch a Mars mission including a rover in 2020. The instruments on board this mission will measure methane concentration in the atmosphere and will probe the ground with a penetrating radar to look for signatures of biological activity.

An artist's view of the Chinese rover to be launched in 2020. *Credit* Xinhua

Another type of mission of major interest is a Mars Sample return mission. The idea is to bring a sample of Martian soil back to Earth, to analyse the sample in well-equipped laboratories on Earth instead of having it analysed automatically in situ. NASA and ESA were discussing plans for such a mission, first separately, then together, then separately again, but as yet no firm dates have been announced. China also has the ambition to set up such a mission by 2030. However, the technical challenges for such missions are enormous and different approaches could be envisioned with no firm decisions being made.

How do we get to Mars and back?
Well, not so easily as it costs enormous amounts of energy. Earth and Mars are on different, more or less circular orbits around the Sun, with Mars being approximately a half time further from the Sun than Earth. So, we need to cross interplanetary voids, and that takes time and energy. The laws of Celestial Mechanics further prevent travel in direct, straight lines and we will have to take orbital paths around the Sun to go from Earth to Mars and then return to Earth. These paths are called transfer orbits, and there are two typical transfer orbits.

The first one is called the Hohmann transfer orbit, named after the German engineer Walter Hohmann, who proposed in 1925 the solution of a half elliptic orbit (2) to transfer between the two (quasi-) circular orbits of Earth (1) and Mars (3).

The orbital manoeuvre to perform the Hohmann transfer uses two engine impulses (Δv) which move a spacecraft onto and off the transfer orbit.

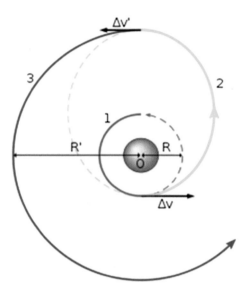

Hohmann transfer orbit. *Credit* Wikimedia Commons

The second transfer orbit, the bi-elliptic transfer orbit, is achieved by two half elliptic orbits where from the initial Earth orbit (in blue), a first engine impulse (at 1) sends the spacecraft into the first transfer orbit (in green) with an aphelion (farthest orbit point from the Sun) at some point (2), where a second engine impulse sends the spacecraft into the second elliptic orbit (in orange) with a perihelion (closest orbit point to the Sun) (3) at the radius of the final desired Mars orbit (in red), where a third engine impulse is performed, injecting the spacecraft into the desired orbit. Despite requiring one more engine burn than a Hohmann transfer and generally requiring a greater travel time, some bi-elliptic transfers require a lower amount of total energy than a Hohmann transfer, that is less propellant to burn in total.

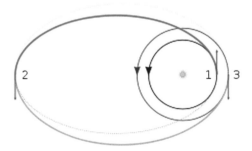

Bi-elliptic transfer orbit. *Credit* Foter.com

These two transfer orbits are those that require the less propellant mass to be used and, depending on mission parameters, one can be more advantageous than the other or vice versa. More detailed information can be found in (Pletser, 2013).

A mission to Mars would require first to examine several aspects in details.

First, should it be an automatic or manned mission? For a manned mission, the transfer time must be minimized to reduce the radiation exposure. For an automatic mission, payload mass must be maximized.

Second, should it be a propelled or ballistic transfer flight? A propelled transfer flight is faster but uses more fuel, i.e. more fuel mass to carry and less available payload mass. A ballistic transfer flight is slower but uses less fuel, i.e. more payload mass is available.

Therefore, for an automatic mission, a ballistic flight with a Hohmann transfer is still the best solution requiring the least amount of fuel.

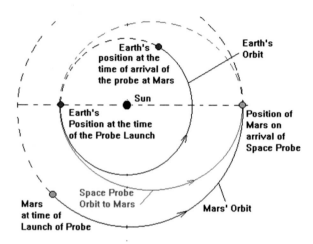

Relative position of Earth and Mars on their orbit around the Sun. *Credit* Geometry in Space

On the above Figure, the perihelion (closest point on the orbit to the Sun) and aphelion (farthest point on the orbit from the Sun) of the Hohmann transfer orbit correspond respectively to the Earth and Mars orbital radius. The duration of the journey Earth-Mars on a Hohmann transfer orbit is 259 days or about 8.5 months, as can be calculated from the laws of celestial mechanics. One can calculate that the launch should take place around 260 days before arrival on Mars and when Mars is approximately 44° ahead of Earth on its heliocentric orbit (see below Figure).

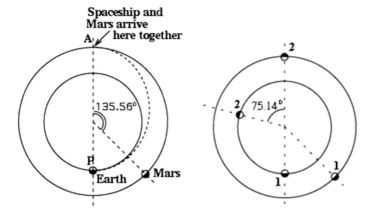

Relative position of Earth and Mars at launch (left) and at arrival (right). *Credit* Stargaze/Smars1 and Smars3

This Hohmann transfer orbit looks promising but has one major drawback: the precise timing for the launch window that occurs about every 26 months.

To return, an automatic mission cannot be launched directly after arrival. How long do we have to wait? It can be calculated that the launch of the automatic return mission should be delayed by 459 days, approximately a year and three months.

A first impulse reduces the spacecraft orbital velocity to enter the return Hohmann transfer orbit and after 260 days, either a second impulse will allow it to enter in Earth's orbit or an atmospheric re-entry will allow an aerobrake to a safe speed for landing back on Earth.

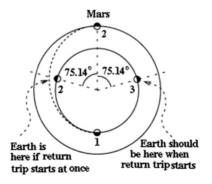

Relative position of Earth and Mars for return. *Credit* Stargaze/Smars3

For a manned mission, the transfer time must be minimized mainly to reduce the exposure time to radiations but also, in case of a ballistic flight, to reduce the debilitating effects of microgravity on astronauts. So, a Hohmann transfer of 8.5 months would not be the best choice.

If the impulse is increased to have transfer orbits with aphelia further and further away and the launch is timed with the predicted positions of Mars, the transfer time can be shortened significantly.

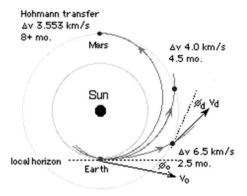

Transfer orbits faster that Hohmann transfer. *Credit* Nordley

Transfer durations decrease obviously with increasing impulse values. The fastest would be for a change of velocity of 6.5 km/s with a transfer time of 76 days. The second braking impulse could be replaced by aerobraking in Mars' atmosphere.

Another option is a constantly propelled transfer flight, accelerating through half of the journey and decelerating for the second half. This would eliminate the debilitating effects of microgravity but at the cost of a larger fuel mass to be embarked.

Instead of classical chemical rockets, another design is being studied and tested, the VASIMR (Variable Specific Impulse Magnetoplasma Rocket) engine. This engine could constantly propel a manned mission to Mars with the advantage of carrying a reduced mass of engine and fuel than a chemical rocket. However, the performances of this type of engine are not yet optimum and its development would probably still require many years.

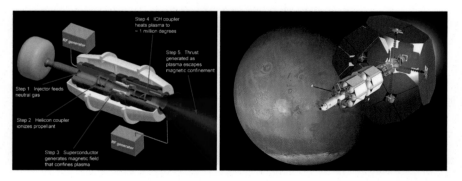

The VASIMR engine concept (left) and artist view of a rocket propelled by a VASIMR engine. *Credit* Ad Astra Rocket Company

There is another possibility, a combination of a ballistic flight, for example with a short Hohmann transfer orbit, and artificial gravity created by centrifugation. A simple design has been proposed by Robert Zubrin, the President of *The Mars Society*. The interplanetary spacecraft is composed of two main vessels that are separated at the beginning of the flight and attached together by a long and thin (but solid enough) tether. The two parts, of more or less equal masses, are then put into rotation by firing small rocket engines. Both then rotate around a point close to the middle of the tether. This dynamic configuration would allow an artificial gravity to be created due to the centrifuge force in both parts of the spacecraft.

So, are we ready to go?

Well, yes and no, there are still some points to clarify. We know already a lot about Mars, and certainly much more than that we knew about the Moon when President J.F. Kennedy, in May 1961, made his call "...*before this decade is out, of landing a man on the Moon and returning him safely to the Earth*". However, the two situations are slightly different. The Moon is not far; it is practically next door: it is less than 400,000 km and it took about two days for the Apollo astronauts to travel to the Moon. The distance between Earth and Mars varies between 56 and 400 million kilometres and as explained above, it will take several months to get to Mars. As demonstrated by the Apollo 13 mission, during which a tank attached to one of the modules exploded on the way to the Moon, the astronauts could not just turn back and come back to Earth; they had to continue, execute half an orbit around the Moon and then come back to Earth, which took them only a couple of days. In the case of a voyage to Mars, this slingshot and return would still take about 500 days in the case of a Hohmann transfer orbit to and from Mars. This time scale and distance are completely different. This means that all aspects and potential risks of a Mars mission must be evaluated, carefully analysed and solutions proposed to reduce the potential risks.

What then are the risks? Let's separate the main issues into two parts: those related to the interplanetary journey and those linked to the stay around and on Mars.

During the journey, we have seen that for a manned mission, a propelled flight or a flight with some kind of artificial gravity would be preferable, for two reasons: to minimize the debilitating effects of microgravity and to minimize the risks associated to radiations exposure.

Let's start with the debilitating effects of microgravity on the human physiology. With the first long duration human space flights at the end of the sixties and early seventies, medical doctors and scientists quickly realised that exposure to microgravity induces different physiological changes, namely bone mass loss due to demineralization; weakening of mass bearing muscles, mainly leg muscles; modifications of the cardiovascular system; changes of the inner ear equilibrium system; and the drastic diminution of the body natural defences, the immune system. Among all these adaptation problems, two problems are really important, namely bone demineralization and decreased immune system; all other problems can be tackled more or less appropriately.

A lot of research has been conducted on astronauts in space and during aircraft parabolic flights. Not only to understand the loss of bone mass and the muscle weakening, but also to try to counteract these effects. But what are the facts? One observes a demineralization, mainly a decalcification, of the bone structure of the astronauts, i.e. a loss of calcium and of phosphorus. Why? One still doesn't know exactly. One could say very simply that the skeleton bones do not have to fight against gravity's pull and do not have to support the weight of the body anymore. Consequently, bones atrophy one way or another and bone cells do not regenerate like on Earth. During Skylab missions in the seventies, it was measured that, on average, approximately 100 milligrams of calcium was lost per day, while an adult organism contains approximately 1 kg of calcium. Bones weaken very rapidly to a point where an astronaut who spent several months in microgravity in orbit or during an interplanetary journey, could not land back on Earth or on another planet, without multiple bone fractures! We know that decalcification is related to an atrophy of the fibrous cells of calcium containing bones, corresponding to the part of the bone that allow the passage of bone marrow. This problem of decalcification resembles in certain aspects to the illness known on Earth as osteoporosis, which affect mainly elderly people and women after menopause. This affliction induces also a change of structure (demineralization) of the bones, associated with the loss of calcium in conjunction with phosphorus, nitrogen and the hydroxyproline protein (another bone component); but the composition stays globally the same. The bone becomes less dense, weakens and fractures more easily. This demineralization is due on one hand to an increased activity of the osteoclastic cells, whose role is to eliminate and resorb bone tissue elements, and on the other hand, to a decreased activity of osteoblastic cells, responsible for the regeneration of bone tissues. To counteract this, space medical doctors prescribe that astronauts partake in daily sports activities, e.g. running on a treadmill for 2 h a day, or working out on an ergometric bicycle or using a squat machine, all of which are integrated in the ISS. It is also a little ironic, when you think about it: the ISS moves at a speed of 25,000 km/h and astronauts are asked to run or pedal on a bicycle, which enables them to actually run faster than Usain Bolt or cycle faster than any *Tour de France* champion.

Regarding the immune system, astronauts' immune defences decrease in microgravity after approximately seven days of flight. One observes a reduction of production of lymphocytes T-cells (the white cells) that intervene in the immune responses and in antibody production. Studies on ISS astronauts showed that some immune cell activity is much lower than normal, while other activity is increased, leaving the Astronauts immune systems completely "confused". A reduced activity may prevent a proper response to germs or viruses' attacks, while an increased activity can induce an excessive response that could yield increased allergy symptoms. Satisfactory explanations are yet to be found for these observations. However, it is believed that many other issues related to spaceflight (radiation, stress, altered sleep cycles, isolation, microbes, …) could affect astronauts' immune systems. Furthermore, astronauts are more prone to infection in space and they need more time to recover after an infection on their return to Earth. The immune defences take five to ten days to return to their normal levels. This problem could be among those that could prohibit mankind in adapting to long duration space travels in microgravity. Let us note however that the interest for this kind of research is presently quite large. Indeed, the understanding of defence mechanisms of the human organism and the role played by gravity, or its absence, can shed light on the fundamental properties of the immune system, of primordial importance in the fight against certain viral affections still incompletely understood on Earth.

Radiation exposure is another issue that may cause significant problems to human crews on their way to, or back from, Mars. Today's astronauts orbiting close to Earth are protected from most space radiation by our planet's upper atmosphere and magnetic field. But heading farther out to space would increase the crews' radiation exposure, from the charged particles expelled by our Sun, known as Solar Particle Events (SPE), and from the heavy ions thrown out by the rest of the cosmos, known as Galactic Cosmic Radiation (GCR). The greatest concern is about the 1% of GCR nuclei the size of an iron atom or more, known as high-ionising high-energy particles (HZE for short) moving at a very high speed and accelerated by magnetic fields across the Universe. HZEs can slice through deoxyribonucleic acid (DNA), a self-replicating material which is present in nearly all living organisms as the main constituent of chromosomes and carries the genetic information. The most serious class of damage is termed 'double-strand breaks', leading to the loss of genetic information and potentially triggering cancer.

The problem is therefore double (SPEs and GCRs) in two different configurations (journey to and from Mars and stay on Mars).

A very interesting piece of information was made available in 2012 by the team taking care of the Radiation Assessment Detector (RAD) instrument aboard NASA's Mars Science Laboratory (MSL) rover Curiosity. It showed that the GCR dose rates on Mars' surface are about half those that RAD measured during its interplanetary cruise.

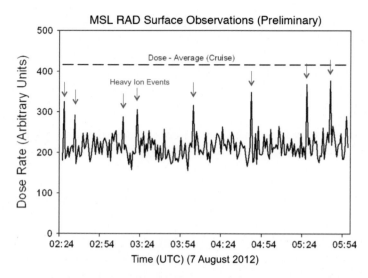

This graphic shows the flux of radiation detected by Curiosity's Radiation Assessment Detector (RAD) on Mars over three and a half hours on 7 Aug. 2012 UTC (Coordinated Universal Time). The data show that the radiation levels measured on Mars during this period of quiet solar activity are reduced from the average radiation detected in space during Curiosity's cruise to Mars. This is explained by the rover being on the planet versus out in space, where it would have more exposure to radiation from all directions. Red arrows point to spikes in the radiation dose rate from heavy ion particles, which would be the most dangerous to astronauts. *Credit* NASA/JPL-CALTECH/ SWRI

Interplanetary GCR dose rates were previously measured by the MARIE instrument aboard the Mars Odyssey spacecraft during its cruise to Mars in 2001 and shown to be about twice that experienced in Low Earth Orbit (LEO). Thus, in combination, the MARIE and RAD results show that Mars surface GCR dose rates are about the same as those experienced by astronauts in LEO.

From this, *The Mars Society* President, Robert Zubrin, concludes that GCR doses will not be a show-stopper for the human exploration of Mars.

However, not all scientists and engineers agree with this conclusion. Intense SPE can cause acute radiation syndrome or radiation sickness, while long term GCR exposure can induce late radiation effects (e.g. cancer).

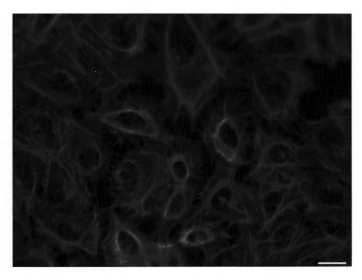

A fluorescent microscopic view of osteosarcoma cells, a type of bone cancer cells, chosen for space experiment because of their rapid growth characteristics. The image has been fluorescently stained to visualise the cells' nuclei in blue and their surrounding cytoskeleton in red. The image covers a length of approximately 305 micrometres (0.305 mm). The scale bar at bottom right measures 25 micrometres across. *Credit* ESA

The space radiation environment, that includes protons and heavy ions, is complex and different from the Earth's natural radiation, mainly α radiation (helium-4 nuclei, with two protons and two neutrons), β radiation (electron and positron, more penetrating than α radiation) and γ radiation (high energy photons, much more penetrating than α and β radiations). Therefore, data taken from the Earth's environment cannot be transposed as such to an interplanetary space environment and the uncertainty in radiation risk estimates is very high.

What are the implications then? Should we completely forget about going to Mars? No, not at all of course. But it does mean that we need to find the proper answer to this problem. Once again, we have several possibilities for the journey to and from Mars.

The first option is to do nothing. It means that we do not take any special measures. Astronauts after all have spent periods of more than one year in space in Earth orbit, with slightly higher risks of cancer than that of normal workers in the nuclear industry on Earth. We could also send astronauts of a certain age and after they had a family and raised their children. Furthermore, choose a period of low Solar activity and a Earth-Mars configuration that would allow a journey as short as possible with a Hohmann transfer orbit, typically 6 months, or a slightly shorter journey with a constant thrust ionic engine. It could work but would still be very risky.

The other two options would combine shielding and shortening the exposure time, i.e. shortening the interplanetary journey.

The second option is to provide some shielding to the astronauts in their spacecraft with materials that could protect them during SPE storms and the GCR background. One would usually think of something like a thick piece of very heavy metal, like lead, but, this is not the best solution as, first, it is not very effective and second, it is prohibitively expensive to launch as the mass density of lead is quite high. A better approach would be to use a lighter material. Liquid hydrogen would do; however, it has to be kept in a liquid state so at very low temperature which would require a very complicated cryogenic plumbing. Hydrogen nanofibers could also do the job. Surprisingly enough, polyethylene plastic would also provide a good shielding. Aluminium also, but not quite as good as polyethylene. Finally, water would provide also an excellent shielding. So, all water tanks could be positioned around the crew capsule and water recycled (like it is done currently now on the ISS) to replenish the protective tanks. Furthermore, one can add a "safe haven" room in the middle of the spacecraft where the crew can take refuge for the two or three days that an intense SPE storm would last. This type of shielding is called passive shielding (for obvious reasons).

The third option is the active shielding where gigantic magnets would surround the spacecraft to deflect all incoming radiations. This is still a science-fiction scenario as the amounts of energy needed to create the deflecting magnetic fields are enormous and not yet feasible. Furthermore, the total mass to launch from ground to orbit and then on a transfer orbit to Mars would render this scenario absolutely unfeasible.

Shortening the interplanetary journey is certainly an interesting option but we do not have yet the engine for this. We talked about the VASIMR engine under development. Another possibility would be nuclear propulsion that could be an interesting option. However, these types of engine have the big disadvantage of not existing yet. We are still many years away from having them built, tested and ready to use for a Mars mission. Maybe one day but not before several decades.

This leaves us with a spacecraft that should include enough passive shielding to protect against the GCR background and with a "safe haven" room for SPE storms. If a Hohmann transfer orbit was chosen, then the spacecraft would need to be put in rotation to generate a centrifugal force that would act as artificial gravity or no rotation at all if a constantly propelled flight approach was chosen.

Now, let us look on the other part, the stay on Mars.

Following the motto put forward by *The Mars Society*, adapted from the first European colonists of the newly discovered America, "Travel light and live off the land", we could use one of the resources on Mars, such as its regolith. Indeed, a certain thickness of Martian sand and rocks would provide a sufficient passive shielding. Furthermore, instead of digging a hole in the ground to bury the first astronauts, let us take advantage of what erosion accomplished naturally on Mars and use those holes on top of lava tubes described earlier as a natural habitat where the first Mars bases could be assembled and deployed. These natural locations could provide shelter, shielding, possibly water as ice if at the right depth, and, who knows, possibly also some forms of ancient or primitive life to study.

Extravehicular activities (or EVA in space-talk) should be conducted to explore the surroundings and a pressurized rover with adequate shielding would be needed.

Many other details and aspects of a future manned mission on Mars are explored in the literature, listed in the reference section.

Another aspect not yet considered concerns the human himself. It is actually the weakest link in a Mars mission but putting a human on Mars remains also the main reason for the mission. Beside all the physiological complications due to the long exposure to weightlessness and to radiations, another aspect concerns the psychological factors. Several issues are at play here: isolation, confinement, boredom, lack of stimulation, constant potential danger, loss of contact with family and friends, group dynamics, etc. Even the most carefully selected and most well trained crew will face interpersonal issues when confined in a small volume with a sense of permanent risks of explosion, depressurization, and radiation to name but a few. So, what can be done to avoid issues such as these becoming potentially fatal for a long duration mission when there is no way out or no possibility of increasing the volume or negating the risks? Well, not much in reality. Once the problem appears, it is usually too late. Problems like these should not appear, simply. First, potential crew members must be carefully selected to apply for a position in a Mars mission crew and second, the whole crew must be well trained to face all potential risks and dangers not as an individual but as a group and react together in harmony. This is of course easier said than done.

Such are the reasons why space agencies and other institutions across the world organized long duration space missions and long duration mission simulations since several decades.

Russian cosmonauts have spent more than two years in space over several flights and one cosmonaut, Valeri Polyakov, still holds the world record for the longest stay in orbit in one go. He stayed 437 days in 1994–95 on board the Russian Mir Station.

Russian Cosmonaut Valeri Polyakov supervised from the Mir module window the rendezvous operations with the Space Shuttle Discovery on its STS-63 in February 1995 (left, *credit* NASA); and acted a subject for a medical experiment operated by Russian Cosmonaut Sergei Krikalev (right, *credit* Roskosmos)

More recently, two astronauts, the American Scott Kelly and the Russian Mikhail Korniyenko, spent 342 days on board the International Space Station during what is known as the "Year Long Mission" in 2015–16. During these long duration missions, it was shown that astronauts and cosmonauts could work and live in space for more than a year and it shows that travels through space from a psychological point of view is possible for a well-trained and well prepared astronaut.

In parallel to these space missions, several ground simulations were also conducted to study more particularly one or several aspects of psychological interactions among crew members isolated from the rest of the world and confined in relatively small spaces for relatively long periods of time. The first simulations were organized by the Russian Institute of Biomedical Problems (IBMP), located in Moscow. Several isolation mission simulations had been already conducted in the seventies and eighties lasting from several days to up to a month. The first major simulation was the SFINCSS-99 (Simulation of Flight of International Crew on Space Station) simulation which took place in Moscow for 8 months in 1999 with a mixed gender crew of several nationalities and which also involved visiting crews in isolation. The next series of long duration isolation missions were also an international effort between Russia, ESA and China and hosted by the Russian IBMP. Named MARS-500, its aim was clearly to prepare for future Mars missions by studying psychosocial interactions between crew members of different cultures and languages. The module assembly was specially prepared for this series of simulations. It included five different modules. A core spacecraft included three modules: the habitat module (with the crew living quarters and 6 small rooms), a medical module (with medical facilities for experiments and telemedicine) and a utility module (with a large refrigerator and storage compartment, a greenhouse, and a bathroom/sauna/gym facility). A fourth module simulated the Mars lander module, connected to the fifth module simulating the Martian surface.

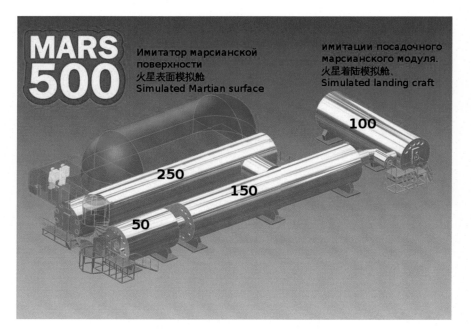

The Mars-500 module complex installed in the IBMP installation with five modules for habitat (150 and 50), medical (100), utility (250), Mars lander and Mars surface. *Credit* IBMP

The simulation took place in three stages. The first stage was a preparatory effort that lasted 15 days with a whole Russian crew of five men and one woman. The second stage involved a male only crew, with four Russians and two Europeans, a German and a French, which lasted 105 days in the first half of 2009. The third round, and the most publicised, lasted 520 days and also consisted of a male only crew, selected among 6000 applicants from 40 countries, and included three Russians, two Europeans, a French and an Italo-Colombian, and a Chinese crew member.

The main goals of this extraordinary simulation were to study the psychophysiological effects of isolation of a confined international six-person crew and to simulate a Mars landing for a sub-crew of three persons. On 3rd June 2010, these six Martionauts boarded the Mars500 facilities, with approximately five tons of food and three tons of water. The door was locked for more than 500 days, more than a year, four months and two weeks. This was truly the first Mars mission and although this mission did not leave Earth, all experiments on board were conducted with the same procedures, the same protocols and the same radio communication delays as for a real Mars mission.

After eighth months of isolation, a subset of three crew members underwent a 30 days hypokinesia (i.e. they stayed in bed for 30 days to simulate the effects of microgravity on the body, mainly on the cardiovascular system). After this microgravity simulation which took place within the main simulation, three simulated Extra-Vehicular Activities (EVA) were performed by two Martionauts in turns in February 2011. After these simulated EVAs, which included a collection of ground samples, they joined again the rest of the crew to continue the isolation simulation for the remaining 240 days.

It would be difficult to summarize the results of the several hundreds of international experiments that were conducted during this simulation. The major outcome of this simulation was to prove that an international crew could conduct successfully a scientific mission of exploration during more than 500 days in perfect harmony. No interpersonal issues were noted among crewmembers and some interesting aspects were observed, such as some crew members experienced perturbed sleeping cycles, while others reported personal mood changes over the entire simulation duration. Technical operational difficulties and simulated emergencies (e.g. a false fire alarm was launched during the night) were faced by the whole crew without any reservation. The crew spent also a lot of time together, not only during meals but also during free time, watching DVDs, preparing surprises for each other, celebrating birthdays and other national celebrations.

All of this proves that selecting the right candidates and following appropriate training and preparation are key to the success of this type of mission.

Other simulations were also conducted throughout the world. *The Mars Society* initiated several simulations in the Western world. After its creation in 1998, it deployed the first Habitat, the FMARS, in the Canadian Arctic (see the first part) in 2000-01 and the second, the MDRS in the Utah Desert, in 2002 (see the second and third parts). At this date (January 2017), more than 170 crews have spent rotations of 15 days on average in MDRS and about 20 crews have spent rotations at FMARS in the Arctic. In September 2015, *The Mars Society* celebrated the thousandth crew member to have stayed at FMARS or MDRS. In Autumn 2016, a crew spent 80 days at MDRS and will spend another 80 days at FMARS, in preparation for a yearlong simulation at FMARS called Mars Arctic 365, to simulate an exploratory mission on Mars.

These simulation activities initiated by *The Mars Society* show the general interest of the public and the specialists, engineers, scientists, journalists, artists and other who have contributed to these missions for more than 15 years.

They have also triggered other simulations at NASA, in laboratories at the Johnson Space Centre, in the field at Rio Tinto in Spain, in Antarctica, on a volcano in Hawaii, and under water. The NASA Extreme Environment Mission Operations (NEEMO) facility is an underwater base in Florida, designed to train astronauts for life in space and used for this kind of simulation.

The HI-SEAS mission, a yearlong simulation ended in August 2016 during which a crew of six spent one year in a dome on the Mauna Loa volcano in Hawaii, in isolation and conducting exploratory EVAs wearing simulated spacesuits. Other simulations at the Mauna Loa mountain were organized with durations of between four and eight months, by the University of Hawaii and supported by NASA.

In China, between June and December 2016, a crew of four volunteers, three men and a woman, conducted a six months' isolation simulation in special facilities in Shenzhen. They were selected from among more than 2000 candidates by the Astronaut Centre of China. The main goals of this simulation were to assess how food, water and oxygen can be used and recycled with the "Controlled Ecological Life Support System" technology inspired by the currently used one in China's Shenzhou spacecraft, and to study the physiological effects of a hermetic environment on humans and the changes to human biological rhythms.

The four Chinese volunteers who underwent the isolation simulation in 2016. *Credit* Xinhua

So, what we did 15 years ago, in the Arctic first and then the Desert of Utah, was certainly valuable. Although it was a small part of a larger body of knowledge and know-how acquired throughout the years by all the crews who spent time in isolation, confinement, and simulation conditions to prepare for some operational aspects of future Mars missions, it was therefore important.

So, with all this, are ready to go to Mars?

Well, nearly. Several manned missions are in the planning. The list of all the various projects that have been announced by NASA or the different US presidential administrations, by the European Space Agency (ESA), and by Russia is long and Wikipedia page on "Human mission to Mars" gives an excellent summary.

However, let us highlight a few. Since the end of 2014, NASA has tested a new capsule, the Orion Multi-Purpose Crew Vehicle (MPCV), which can accommodate four astronauts for launch from Earth, re-entry and splashdown into the ocean, similar to an Apollo capsule but slightly bigger. This capsule is complemented by a Service Module developed by ESA and based on the design of ESA's Automated Transfer Vehicle that was used five times to automatically resupply the ISS.

An artist's impression of the Orion spacecraft with ESA's service module. The module sits directly below Orion's crew capsule and provides propulsion, power, thermal control, and water and air for four astronauts. The solar array spans 19 m and provides enough power to power two households. A little over 5 m in diameter and 4 m high, it weighs 13.5 tonnes. The 8.6 tonnes of propellant will power one main engine and 32 smaller thrusters. *Credit* ESA—D. Ducros

The first unmanned launch of the Orion capsule took place in December 2014. In March 2017, NASA announced its present plan to send humans to Mars, that includes five phases, Phases 0 to IV. Phase 0 is already underway since sometime and it includes using the ISS for preparatory work and testing along with discussions with private space companies and international partners. During Phase I, planned from 2018 till 2025, six heavy expendable Space Launch System (SLS) rockets are planned to be launched and tested. An automatic mission is presently planned firstly for 2018, with ESA's Service Module. Throughout the years, these SLS rockets with Orion capsules will deliver the elements and components of a new space station called Deep Space Gateway (DSG) that will be assembled in orbit around the Moon. Phase II, foreseen to start in 2025, will see the development of a new vehicle, called the Deep Space Transport (DST) that would accommodate a crew of four astronauts for up to 1000 days and support three round trips to Mars. This new DST vehicle would be launched in 2027 to the new Moon orbital DSG space station with astronauts on board. Later on, around 2028 and 2029, astronauts would stay in space for long duration missions, up to 400 days, in order to support commercial lunar activities and to prepare future mission to Mars.

The Deep Space Transport (DST) vehicle (right) approaches the future Moon Deep Space Gateway (DSG) space station (left). *Credit* NASA

Phase III, starting in 2030, would allow to configure the DST vehicle from the DSG station with supplies for long duration mission to Mars and the mission to Mars itself is planned in Phase IV, foreseen in 2033.

An ambitious plan that presently lacks two things: a staged budget and a realistic schedule as the latter looks quite tight, especially for Phases II and III. Besides that, it is a good news that NASA now has a plan to put humans on Mars.

Collaboration between ESA and NASA has seen the construction of Orion's Service Module. In addition, ESA has pushed the idea of establishing a global Moon village, but as yet has no clear budget or plans to build it. And for Mars, beside the automatic ExoMars mission in 2020 and a possible participation in a Mars Sample Return mission, ESA has no plans to send humans to Mars, except maybe while cooperating with NASA.

China plans to send an automatic rover to Mars' surface in 2020 with a newly tested Long March 5 rocket. China has the long-term plan to send human crews to Mars within a 2040–2050 timeframe.

Russia has also announced a long-term plan to send manned crews to Mars after 2040.

Several private missions have also been announced.

Denis Tito, the first private astronaut on the ISS, announced in 2013 his plan, called *Inspiration Mars*, to put an American married couple in 2018 on a flyby mission to Mars, i.e. to fly to Mars and return directly without attempting to land. The flight was scheduled to last approximately 500 days. However, not enough funding was raised and other than some polite initial discussions, NASA did not appear to show any interest. It seems that the project is now defunct.

MarsOne is another private effort to put humans on Mars this time, however the approach has appeared unprofessional, despite desperate efforts to present it as serious. In a nutshell, a group of Dutch entrepreneurs, which included only one engineer, announced in 2012 its proposal to launch non-return missions to Mars. After having launch an armada of automatic rovers that would scout the entire surface of Mars for the best spots to deploy the first human colonies in 2024, crews of colonists would be launched to Mars on a one-way trip. Funding initially would have come from rights sold to television companies specialized in TV reality shows. This funding aspect already raised suspicion about the seriousness of the project. Such a program would first raise serious ethical concerns, and second, from a practical point of view, it could never raise enough monies to pay for itself. Furthermore, the project self-financed itself through its call for astronaut candidates which demanded no particular requirements other than candidates having to send a short video recording of themselves explaining why they wanted to go to Mars and also paying 50 dollars or Euros. Probably the most interesting part of this project was that approximately 220,000 people, the majority with no particular astronaut or space related skills, were ready to embark on a one-way suicide mission to an unknown place. Sociologically, it would be interesting to evaluate how many of these people had had enough of their life on Earth, being ready to leave their family, spouse and children to embark on such a one-way mission (not paying attention to the lack of mission definition). Practically, it means that *MarsOne* received more than 10 million of dollars or Euros to prepare for … What exactly? Nobody really knows as there was never any serious plan to design and build such a mission: no long-term budget, no launcher, no vehicle, no automatic rovers, no habitat for colonists, no plan, no design of colony hardware, etc. Talks were initiated with big space companies and were all received politely and further ignored. Recently, the initiator of the project recognized that their 12-year plan for landing humans on Mars by 2027 was mostly fiction. In the meantime, the selection process of so-called "*MarsOne* astronauts" continued and is now down to 100 candidates. What can be said? All these 220,000 applicants were full of hopes and dreams and these entrepreneurs abused these dreams and hopes. *MarsOne* is exactly what space exploration does not need: at worst, a scam; at best, an amateur enterprise that could completely disgust the general public from space exploration and from genuine efforts of the scientific community. Imagine just an instant that they would gather marginally enough money to launch a first crew with limited resources and, once arrived, to live from Mars resources. How long would it take for this untrained crew to fight among themselves for these limited resources: food, water, air… And all this in front of cameras for the entire world to watch. It is of no surprise that nobody in the space scientific and technical circles supported such an approach. No, this is not how space exploration should be conducted.

A third private effort, more serious this time, was started in 2002 by entrepreneur Elon Musk who created *SpaceX*, an American aerospace company. *SpaceX*'s goals are to develop space transportation technologies at lower costs, to enable the eventual colonization of Mars. *SpaceX* developed several kinds of rockets that captured an important part of the launch market in the world, in competition with the giant companies from the USA (Boeing, Lockheed, …) and Europe (Arianespace). They succeeded in obtaining several launch contracts from NASA to resupply the ISS with their Dragon capsule. *SpaceX* has also developed the technique of landing back vertically the first stage rocket near the launch pad instead of discarding it in the ocean after launch. By reusing the same hardware, this approach has drastically reduced the launch costs.

In September 2016, at the International Astronautical Congress (IAC) in Guadalajara, Mexico, *SpaceX*'s Chief Executive Officer Elon Musk unveiled his plans later on confirmed at the IAC congress in Adelaide, Australia, in September 2017, to develop an Interplanetary Transport System programme, that includes a new launch vehicle, a new spacecraft, and a new mission architecture. This programme would be funded completely privately and could probably be the best chance that mankind has in the coming years to see interplanetary space flights that could lead eventually to sustainable human settlements on Mars.

The concept is relatively simple. The rocket, the *SpaceX* Interplanetary Transportation System, is launched from Florida, with a crewed spacecraft. The spacecraft separates from the rocket and stays in a parking orbit around the Earth. The rocket returns to Earth to land vertically on the launch pad, where a second cargo craft with fuel is installed on the rocket. The fuel cargo craft is launched and rendezvous with the first crewed spacecraft. After docking, fuel is transferred to the crewed spacecraft, which can then leave Earth's orbit, while the empty cargo craft return to Earth. The crewed spacecraft flies to Mars and enters in Mars atmosphere as a winged body, aerobraking on Mars thin atmosphere. It eventually brakes further with its retrorockets and lands vertically on Mars's surface.

From left to right and up to down: series of screen captures of SpaceX animation video showing the launch of the crewed spacecraft, the refuelling in orbit, the solar panels deployment, the Mars atmospheric braking, the Mars landing and the astronauts ready for a first EVA. *Credit SpaceX*

Et voilà ! As simple as that. Well, yes and no. The concept is very simple and individual technology is already existing or nearly ready. The foreseen engine, the Raptor engine is under development and forty to fifty of these engines need to be assembled at the base of this giant rocket of 120 m high, just taller than Apollo's Saturn V. Both crew and cargo vehicles need to be developed with the huge solar panels and all on board subsystems. Then the overall systems need to be integrated, assembled, tested and tried out in real conditions. A little more work yet to be done. The date announced by Elon Musk for a first test flight in 2018 is not really credible and could be pushed most likely to 2020, while a first manned flight could take place in the first half of the 2020s. Beside technical hurdles still to be solved, another concern is budget. How are all of these developments to be funded?

So, we have seen that we are still relatively far from flying a manned crew to Mars in the coming years. Although we have everything at our disposal: technologies exist, know-how is there and knowledge has been acquired. What is missing is a political will to really get on with it. What could really kick the start of this next human adventure in space exploration is two things. First, one discovers irrevocably that life exists or existed in the past and that this life is or was completely different from the one we know on Earth. This would be a new Copernican revolution that would trigger a rush to study this new life form in situ and to install research stations on Mars. Second, one discovers on Mars some kind of ore or element badly needed on Earth, either because of exorbitant production costs or because its rarity on Earth. One or the other or both could help to commit public and/or private funding to be invested in manned Mars missions.

To conclude, Mars while being very close is still quite far away. Very close, because we know already so much about it and we are longing to stroll along ridges and chasms on this new planet. A planet which has a total surface as large as the whole surface of Earth's continents, is there, waiting for us with all its promises of new discoveries and a potential future for a new branch of humanity. And very far, because there is still so much that we do not know about ourselves, how would we react to flying in weightlessness for so long, how would we protect ourselves against radiations, how would we behave on this new planet, would we treat it ethically and with respect or continue to abuse this new environment like we do on Earth.

The two modules in the Arctic and in the desert are still in use for new simulation campaigns, longer and allowing more scientific and exploratory experiments to be conducted. If you are interested, *The Mars Society* recruits volunteers to participate in these simulations. Have a look at their web site and you will find all the information and conditions to participate.

Who knows, among you who are reading these lines may be the first humans to set foot on Mars in twenty or thirty years …

To finish, I invite you to a little guessing game. Imagine that you are reading the headlines from future newspapers on Earth. Can you put a date [dd/mm/yyy] next to each title?

The first human mission to Mars has taken off successfully today. On-board, a mixed gender crew of three men and three women representing six different nationalities, languages and cultures. An interplanetary travel of six months will bring them to our neighbouring planet [dd/mm/yyy]

The first crewed Martian craft landed safely on the surface of the Red Planet. [dd/mm/yyy]

Two astronauts, a Chinese woman and an American man, set foot together on the soil of Mars for the first time in the history of mankind and started the first Martian EVA to deploy scientific equipment. [dd/mm/yyy]

Rover expeditions have now become routine on Mars where the six human explorers take turn in exploring our sister planet. [dd/mm/yyy]

The German biologist of the Martian crew has reported that past evidence of life has been discovered using the PCR method in samples collected in a Martian cave. Could it be a sign? [dd/mm/yyy]

We are not alone! The World Space Agency has confirmed what has already been suspected for some time. A living colony of a new sort of bacteria has been discovered in a deep cave, formed by lava, deep under the surface of Mars. [dd/mm/yyy]

The methane mystery solved! Methane concentration has been measured in the cave where this new bacteria colony has been detected. Findings correlate with measurements of atmospheric concentration made by Mars orbiting satellites. Methane is indeed generated by life forms on Mars. [dd/mm/yyy]

The power generator deployed by the second Martian crew has reached the levels needed for producing enough power to sustain the first future human colony on Mars. [dd/mm/yyy]

The first human baby conceived on another planet than Earth has been born today: it is a girl named Eva. The baby and her mother, a Japanese astrophysicist, are well. Her father, a Russian geologist, is very proud. [dd/mm/yyy]

Terraforming of Mars is proceeding well: the Martian atmospheric pressure is increasing rapidly and the first liquid water ponds are forming naturally. [dd/mm/yyy]

Mars Terraforming is near completion! The Mars atmosphere is nearly breathable with oxygen concentration increasing steadily, all made possible by the vegetation that has adapted to the supplemented Martian conditions of enriched regolith and free flowing water. [dd/mm/yyy]

Earth officials and representatives of the various Mars colonies are meeting today to discuss modalities of commercial cooperation and pertinent political issues [dd/mm/yyy]

Will you (and I) still be alive to see all this? Is it really so difficult?
On to Mars!

Vladimir Pletser
31 January 2017

Technology and Engineering Centre for Space Utilization
Chinese Academy of Sciences
Vladimir.Pletser@csu.ac.cn

References

On Part One The Arctic

The first Mars simulation campaign on websites:

- general presentation:

    http://arctic.marssociety.org/

- the NASA/SETI Haughton-Mars Project on Devon Island:

    http://www.arctic-mars.org

- the crew members of the campaign:

    http://www.arctic-mars.org/team/2000/index.html

- excerpts of the author's diary on the ESA web site:

    http://www.esa.int/Our_Activities/Human_Spaceflight/Exploration/Postcard_from_Mars (or go to the site www.esa.int and search for the link '*Postcards from Mars*')

The first Mars simulation campaign in the international press:

- the Hab deployment:
  "North to Mars", R. Zubrin, *Scientific American*, June 2001, 50–53.
  "Martionautes dans le Grand Nord" (in French), J. Lopez, *Science et Vie Junior*, Dossier Hors-Série No 40, April 2000, 34–43.

- the call for volunteers for this simulation:
  "Mars Society seeks volunteers to perform research in Arctic", B. Berger, *Space News*, 11 December 2000, p. 8.

- the Belgian written press reports on this first campaign in the Arctic:
  "2001 Mars Odyssey: l'aventure recommence—Un Belge 'sur Mars'" (in French), C. Du Brulle, *Le Soir*, 9 April 2001, p. 7.

© Springer Nature Singapore Pte Ltd. 2018
V. Pletser, *On To Mars!*, https://doi.org/10.1007/978-981-10-7030-3

"Mission martienne pour un Belge" (in French), J. Miralles, *La Dernière Heure*, 29 June 2001.

"Vladimir Pletser: Brusselaar bereidt Marsmissie voor tussen ijsberen" (in Flemish), *Nieuwsblad*, 29 June 2001, p. 21.

"Un Belge sur Mars pendant dix jours" (in French), C. Du Brulle, *Le Soir*, 12 July 2001, p. 14.

"Belgische onderzoeker in simulatie van Mars" (in Flemish), *Het Laatste Nieuws*, 13 July 2001, p. 9.

"Sale temps pour l'équipage martien posé dans le Grand Nord canadien" (in French), C. Du Brulle, *Le Soir*, 19 July 2001, p. 12.

"Mission martienne pour Vladimir Pletser" (in French), T. Pirard, *Athéna*, No 174, October 2001, 100–101.

- the international written press reported also on this first campaign. Among other:
  "Let's play astronauts—Why would a group of people spend weeks in an Arctic crater, cooped up in a tin can on stilts?", S. Armstrong, *New Scientist*, 21 July 2001, p. 11

  "Uma experiência no Arico" (in Portuguese), T. Russomano, *Diario Popular*, Porto Alegre, Brazil, 9 September 2001, p. 10.

  "Mars boot camp", F. Vizard, *Popular Science*, October 2001, 52–58.

  "Simulation of a manned Martian mission in the Arctic Circle", V. Pletser, *ESA Bulletin* No 108, November 2001, 121–123.

- Robert Zubrin published his diary in:
  "Dispatches from the Flashline Mars Arctic research Station: Learning how to explore Mars in the Canadian Arctic", R. Zubrin, paper IAA-01-IAA.13.3.09, *52nd International Astronautical Congress*, 1–5 October 2001, Toulouse.

- Charles Frankel, a French crewmember of the third rotation, published his diary in:
  "Chroniques d'un Martien" (in French), C. Frankel, a series of seven papers published in *Libération*, Paris, from 20 to 27 July 2001, p. 16.

  "J'ai vécu dix jours sur Mars" (in French), C. Frankel, *Ciel et Espace*, October 2001, 34–38.

Some technical and scientific papers on results of the seismic experiments:

"Subsurface water detection on Mars by astronauts using a seismic refraction method: tests during a manned Mars mission simulation", V. Pletser, P. Lognonné, M. Diament, V. Dehant *Acta Astronautica* 64 (IF 0.609), 457–466, 2008.
http://www.sciencedirect.com/science/article/pii/S0094576508002609

"Simulation of a manned Martian mission in the Arctic Circle", V. Pletser, *ESA Bulletin* 108, 121–123, 2001.
http://www.esa.int/esapub/bulletin/bullet108/chapter15_bul108.pdf

"Subsurface water detection on Mars by active seismology: simulation at the Mars Society Arctic Research Station", V. Pletser, P. Lognonné, M. Diament, V. Ballu, V. Dehant, P. Lee, R. Zubrin, *Proceedings Conference on Geophysical Detection of Subsurface Water on Mars*, Abstract 7018, Lunar and Planetary Institute, Houston, 6–10 August 2001.

"Feasibility of an active seismology method to detect subsurface water on Mars by a human crew: Summer 2001 Flashline M.A.R.S. campaign first results and lessons learned", V. Pletser, P. Lognonné, M. Diament, V. Dehant, K. Quinn, R. Zubrin, P. Lee, *Proceedings Fourth International Mars Society Convention*, University of Stanford, USA, 23–26 August 2001.

"How astronauts would conduct a seismic experiment on the planet Mars", V. Pletser, P. Lognonné, V. Dehant, *Proceedings of the Symposium on the Influence of Geophysics, Time and Space Reference Frames on Earth Rotation Studies*, Paris Observatory -Belgian Royal Observatory, Brussels, 24–26 September 2001, 147–156.

"Simulation of Martian EVA at the Mars Society Arctic Research Station", V. Pletser, R. Zubrin, K. Quinn, *53rd International Astronautical Congress—The World Space Congress*, Houston, paper IAC-02-IAA.10.1.07, October 2002.

Canada and Nunavut:

"The rough guide to Canada", T. Jepson, P. Lee, T. Smith, *Rough Guides*, Penguin, London, June 2001.

"The Nunavut handbook", M. Soublière, *Nortext Multimedia Inc*, Iqaluit, 1998.

On Part Two The Desert

The second Mars simulation campaign on websites:

- presentation of the crew:

  https://web.archive.org/web/20060927073551/http://www.marssociety.org/mdrs/fs01/crew05/

- diary and scientific reports of our crew:

  https://web.archive.org/web/20060623114408/http://www.marssociety.org/mdrs/fs01/ (and select entries for Crew 5)

- excerpts of the author's diary:
  in English on the ESA web site:

  http://www.esa.int/export/esaHS/ESALDSF18ZC_future_0.html
  (or go to the site www.esa.int and search for the link '*Postcards from Mars 2*')
  in French on the web site of the Belgian newspaper 'La Dernière Heure':

  http://www.dhnet.be/actu/societe/objectif-mars-51b7d439e4b0de6db9909bff and following

The second Mars simulation campaign in the international press:

"Le journal d'un Martien" (in French), excerpts of the author's diary in a series of
    papers published in the '*La Dernière Heure*' newspaper between 8 and 18 April
    2002.
"Peterchens Marsfahrt" (in German), *Die Zeit*, Nr 17, 18 April 2002, p. 31.
"Life on Mars", C. Laurence, *The Sunday Telegraph*, Review, London, 5 May
    2002, p. 1–2.
"Life on Mars", C. Laurence, *Irish Independent*, Weekend, Dublin, 18 May 2002,
    12–14.
"Mission to … Utah?—Would be voyagers to Mars look for lessons about how
    people would fare on a real mission to the fourth planet", D. Real, *The Dallas
    Morning News*, Section C, 6 July 2002, p. 1C–2C.
"Nos vemos en Marte - Una colonia de cientificos se entrena en el desierto de
    Nevada para vivir en Marte" (in Spanish), C. Laurence, *El Pais Semanal*, No
    1349, 4 August 2002, p. 1 + pp. 30–37.

Some technical and scientific papers on our plant growing experiment:

"Would the first astronauts on Mars grow vegetables for their consumption or for
    their psychological well-being?", V. Pletser, C. Lasseur, *Proceedings of the
    Second European Mars Convention EMC2*, Rotterdam, 27–29 September 2002.
"A closed Mars Analog simulation: the approach of Crew 5 at the Mars Desert
    Research Station, April 8–20, 2002", W.J. Clancey, *Proceedings of the Fifth
    International Mars Society Convention*, University of Colorado, Boulder, USA,
    8–11 August 2002.
"First observation regarding the psychological impact of growing vegetables during
    a manned Mars mission simulation at the Mars Desert Research Station", V.
    Pletser, C. Lasseur, *54th Congress International Astronautical Federation*,
    Bremen, paper IAC-03-IAA.10.3.04, October 2003.

On Part Three The Desert Reload

Some technical and scientific papers on our experiments:

"Logbook for day 283 on Mars; Crew 1, Crew Biologist Cora S. Thiel Reporting",
    Thiel C.S., Pletser V., in *One Way Mission to Mars*, Davies P. and Schulze-
    Makuch D. eds, Vol. 13, 4121–4130, January 2011. http://journalofcosmology.
    com/Mars151.html
"A Mars Human Habitat: European approaches and recommendations on crew time
    utilisation and habitat interfaces", Pletser V., in *The Human Mission to Mars.
    Colonizing the Red Planet*, Levin J.S. and Schild R.E. eds, ISBN:
    9780982955239, ISBN-10: 0982955235, 757–784, Oct.-Nov. 2010.
    http://journalofcosmology.com/Mars123.html

"PCR-based analysis of microbial communities during the EuroGeoMars campaign at Mars Desert Research Station, Utah", Thiel C., Ehrengreund P., Foing B., Pletser V., Ullrich O., *Int. Journal of Astrobiology*, 10, 177–190, 2011. doi:10.1017/S1473550411000073; http://journals.cambridge.org/repo_A82FegRf http://journals.cambridge.org/action/displayAbstract_S1473550411000073 https://www.cambridge.org/core/journals/international-journal-of-astrobiology/article/pcr-based-analysis-of-microbial-communities-during-the-eurogeomars-campaign-at-mars-desert-research-station-utah/E50A122EA6EC5DEB3EA7A482B6BE635A

"Human crew related aspects for astrobiology research", Thiel C.S., Pletser V., Foing B. and the EuroGeoMars Team, *Int. Journal of Astrobiology*, 10, 255–267, 2011. doi:10.1017/S1473550411000152; http://journals.cambridge.org/repo_A82BxBSF; http://journals.cambridge.org/abstract_S1473550411000152 https://www.cambridge.org/core/journals/international-journal-of-astrobiology/article/human-crew-related-aspects-for-astrobiology-research/D1887837E0551 FC56A0E7133CAC0E53C

"Field astrobiology research in Moon-Mars analogue environment: instruments and methods", Foing B.H., Stoker C., Zavaleta J., Ehrengreund P., Thiel C., Sarrazin P., Blake D., Page J., Pletser V., Hendrikse J., Dirieto S., Kotler M., Martins Z., Orzechowska G., Grozz C., Wendt L., Clarke J., Borst A., Peters S., Wilhelm M.B., Davies G, and ILEWG EuroGeoMars 2009 support team, *Int. Journal of Astrobiology*, 10, 141–160, 2011. doi:10.1017/S1473550411000036 http://journals.cambridge.org/repo_A82y5h00 http://journals.cambridge.org/action/displayAbstract?fromPage=online&aid=8286802&fulltextType=RA&fileId=S1473550411000036 http://journals.cambridge.org/action/displayAbstract?fromPage=online&aid=8286802 https://www.cambridge.org/core/journals/international-journal-of-astrobiology/article/field-astrobiology-research-in-moonmars-analogue-environments-instruments-and-methods/89160676373253E16D32C1412C128794

"European contribution to human aspect field investigation for future planetary habitat definition studies: field tests at MDRS on crew time utilization and habitat interfaces", Pletser V., Foing B., *Microgravity Science and Technology*, 23–2, 199–214, 2011. https://doi.org/10.1007/s12217-010-9251-4 http://www.springerlink.com/content/a820483350l2jh78/

"A Mars Human Habitat: European approaches and recommendations on crew time utilisation and habitat interfaces", Pletser V., *Journal of Cosmology*, Special Issue on 'The Human Mission to Mars: Colonizing the Red Planet', Vol. 12, Oct.–Nov., 2010, 3928–3945. http://journalofcosmology.com/Mars123.html

"Preliminary lessons learnt after the rotation of the first EuroGeoMars team—Crew 76", Pletser V., Technical Report HSF-UP/2009/113/VP, ESTEC, Noordwijk, 2009.

"Field reports of science and technology activities of the first EuroGeoMars team—Crew 76 at the Mars Desert Research Station, 1–14 February 2009", Pletser V., Monaghan E., Peters S., Borst A., Wills D., Hendrikse J., *Report MDRS-76/FR-01*, Mars Desert Research Station, The Mars Society, 2009.

"Lunar outpost pre-design: Human aspects", Boche-Sauvan L., *Master Project BO-F09003*, Arts et Métiers ParisTech—GeorgiaTech, 2009.

"Human interfaces study: Framework and first results", Boche-Sauvan L., Pletser V., Foing B.H., EuroGeoMars Crew, *European Geosciences Union, General Assembly 2009—Vienna*, 2009.

"Human aspects study through industrial methods during an MDRS mission", Boche-Sauvan L., Pletser V., Foing B.H., EuroGeoMars Crew, *NASA Lunar Science Institute—Ames Research Center*, 2009.

"Human aspects and habitat studies from EuroGeoMars campaign", Boche-Sauvan L., Pletser V., Foing B.H., ExoGeoMars team, *EGU2009-13323*, Geophysical Research Abstracts, Vol. 11, 2009.

"Geochemistry of Utah Morrison formation from EuroGeoMars campaign", Borst A., Peters S., Foing B.H., Stoker C., Wendt L., Gross C., Zhavaleta J., Sarrazin P., Blake D., Ehrenfreund P., Boche-Sauvan L., Page J., McKay C., Batenburg P., Drijkoningen G., Slob E., Poulakis P., Visentin G., Noroozi A., Gill E., Guglielmi M., Freire M., Walker R., Sabbatini M., Pletser V., Monaghan E., Ernst R., Oosthoek J., Mahapatra P., Wills D., Thiel C., Lebreton J.P., Zegers T., Chicarro A., Koschny D., Vago J., Svedhem H., Davies G., Westenberg A., Edwards J., ExoGeoLab team & EuroGeoMars team, *Int. Conf. Comparative Planetology: Venus—Earth—Mars*, ESTEC, Noordwijk, 2009.

"Terrestrial field research on organics and biomolecules at Mars Analog sites", Ehrenfreund P., Quinn R., Martins Z., Sephton M., Peeters Z., van Sluis K., Foing B.H., Orzechowska G., Becker L., Brucato J., Grunthaner F., Gross C., Thiel C., Wendt L.: *Int. Conf. Comparative Planetology: Venus—Earth—Mars*, ESTEC, Noordwijk, 2009.

"Geology and geochemistry highlights from EuroGeomars MDRS campaign", Foing B.H., Peters S., Borst A., Wendt L., Gross C., Stoker C., Zhavaleta J., Sarrazin P., Blake D., Ehrenfreund P., Boche-Sauvan L., Page J., McKay C., Batenburg P., Drijkoningen G., Slob E., Poulakis P., Visentin G., Noroozi A., Gill E., Guglielmi M., Freire M., Walker R., Pletser V., Monaghan E., Ernst R., Oosthoek J., Mahapatra P., Wills D., Thiel C., Lebreton J.P., Zegers T., Chicarro A., Koschny D., Vago J., Svedhem H., Davies G., Westenberg A., Edwards J., ExoGeoLab team & EuroGeoMars team, *Int. Conf. Comparative Planetology: Venus—Earth—Mars*, ESTEC, Noordwijk, 2009.

"ExoGeoLab lander and rover instruments", Foing B.H., Page J., Poulakis P., Visentin G., Noroozi A., Gill E., Batenburg P., Drijkoningen G., Slob E., Guglielmi M., Freire M., Walker R., Sabbatini M., Pletser V., Monaghan E., Boche-Sauvan L., Ernst R., Oosthoek J., Peters S., Borst A., Mahapatra P., Wills D., Thiel C., Wendt L., Gross C., Lebreton J.P., Zegers T., Stoker C., Zhavaleta J., Sarrazin P., Blake C., McKay C., Ehrenfreund P., Chicarro A.,

Koschny D., Vago J., Svedhem H., Davies G., ExoGeoLab team & EuroGeoMars team, *Int. Conf. Comparative Planetology: Venus—Earth—Mars*, ESTEC, Noordwijk, 2009.

"Highlights from Remote Controlled Rover for EuroGeoMars MDRS Campaign", Hendrikse J., Foing B.H., Stoker C., Zavaleta J., Selch F., Ehrenfreund P., Wendt L., Gross C., Thiel C., Peters S., Borst A., Sarrazin P., Blake D., Boche-Sauvan L., Page J., Pletser V., Monaghan E., Mahapatra P., Wills D., McKay C., Davies G., van Westrenen W., Batenburg P., Drijkoningen G., Slob E., Poulakis P., Visentin G., Noroozi A., Gill E., Guglielmi M., Freire M., Walker R., ExoGeoLab team & EuroGeoMars team, *European Planetary Science Congress*, EPSC Abstracts, Vol. 4, 2009.

"A Prototype Instrumentation System for Rover-Based Planetary Geology", Mahapatra P., Foing B., Nijman F., Page J., Noroozi A., ExoGeoLab team, *European Planetary Science Congress*, EPSC Abstracts, Vol. 4, 2009.

"Alluvial fan EuroGeoMars observations and GPR measurements", Peters S., Borst A., Foing B.H., Stoker C., Kim S., Wendt L., Gross C., Zhavaleta J., Sarrazin P., Blake D., Ehrenfreund P., Boche-Sauvan L., Page J., McKay C., Batenburg P., Drijkoningen G., Slob E., Poulakis P., Visentin G., Noroozi A., Gill E., Guglielmi M., Freire M., Walker R., Sabbatini M., Pletser V., Monaghan E., Ernst R., Oosthoek J., Mahapatra P., Wills D., Thiel C., Petrova D., Lebreton J.P., Zegers T., Chicarro A., Koschny D., Vago J., Svedhem H., Davies G., Westenberg A., Edwards J., ExoGeoLab team and EuroGeoMars team, *Int. Conf. Comparative Planetology: Venus—Earth—Mars*, ESTEC, Noordwijk, 2009.

"Basic Mars Navigation System For Local Areas", Petitfils E-A., Boche-Sauvan L., Foing B.H., Monaghan E., EuroGeoMars Crews: *EGU2009-13242-2*, Geophysical Research Abstracts, Vol. 11, EGU General Assembly, 2009.

Other technical and scientific papers:

"Participant Observation of a Mars Surface Habitat Mission Simulation", Clancey W.J., *Habitation*, 11(1/2) 27–47, 2006.

"HUMEX, a study on the survivability and adaptation of humans to long-duration exploratory missions, part II: Missions to Mars", Horneck G., Facius R., Reichert M., Rettberg P., Seboldt W., Manzey D., Comet B., Maillet A., Preiss H., Schauer L., Dussap C.G., Poughon L., Belyavin A., Reitz G., Baumstark-Khan C., Gerzer R., *Adv. Space Res.* 38–4, 752–759, 2006.

"Exploration of Mars: the reference mission of the NASA Mars exploration study team", Hoffman S., Kaplan D., *NASA SP 6107*, Houston, 1997.

On Part Four Mars Tomorrow

Mars missions on websites:
ESA website

>   http://www.esa.int/Our_Activities/Space_Science/Mars_Express/Mars_Express_
>   mission_facts

NASA website

>   http://mars.jpl.nasa.gov/msl/

Methane in Mars atmosphere
http://exploration.esa.int/mars/46038-methane-on-mars/
https://www.nasa.gov/mission_pages/mars/news/marsmethane.html

Holes on Mars
http://www.lpi.usra.edu/meetings/lpsc2007/pdf/1371.pdf

"Face" on Mars

- Proponent books
  "Planetary Mysteries: Megaliths, Glaciers, the Face on Mars and Aboriginal
  Dreamtime", Grossinger R., ed., Berkeley, North Atlantic Books, p. 11, 1986.
  ISBN 0-938190-90-3.
  "The Monuments of Mars: A City on the Edge of Forever", Hoagland R., North
  Atlantic Books, USA, 2002. ISBN 978-1-58394-054-9
- Rebuke of Carl Sagan
  "The Demon-Haunted World: Science As a Candle in the Dark", Sagan C.,
  Random House, New York, 1995. ISBN 978-0-394-53512-8.

China Mars mission 2020
http://www.chinadaily.com.cn/china/2016twosession/2016-03/05/content_
    23747640.htm
http://www.scmp.com/news/china/society/article/1902837/chinas-first-mission-mars-
    will-be-hugely-ambitious-and-be-chance
http://www.bbc.com/news/av/world-asia-36085659/when-will-china-get-to-mars
http://www.dailymail.co.uk/sciencetech/article-3549536/China-wants-land-Mars-
    2021-official-country-s-space-agency-reveals-plans-mission-red-planet.html
https://www.youtube.com/watch?v=XiJE5x9Lc80

How to get to Mars?

"Mathematics In Space: Free-Falling, The Way Home And Far Away…(Part 2)", Pletser V., invited plenary lecture, *Proc. 16th Biennial Congr. Flemish Mathematics Teachers Association VVWL* 2012, Blankenberge, Belgium, July 2012; Wiskunde and Onderwijs, VVWL Tijdschrift, ISSN 2032-0485, Nr 155, 230–246, 2013.

https://www.researchgate.net/publication/312086560_Mathematics_In_Space_Free-Falling_The_Way_Home_And_Far_AwayPart_2

http://foter.com/photo/bi-elliptic-transfer/

http://math-ed.com/Resources/GIS/Geometry_In_Space/OrbitalTransfer.htm

http://www.phy6.org/stargaze/Smars1.htm

http://www.phy6.org/stargaze/Smars3.htm

G. Nordley, Going to Mars?, 2006, http://www.gdnordley.com/_files/Going_to_Mars.html

"VASIMR Performance Measurements at Powers Exceeding 50 kW and Lunar Robotic Mission Applications", J. P. Squire, F. R. Chang-diaz, T. W. Glover, M. D. Carter, L. D. Cassady, W. J. Chancery, V. T. Jacobson, G. E. Mccaskill, C. S. Olsen, E. A. Bering, M. S. Brukardt, B. W. Longmier, International Interdisciplinary Symposium on Gaseous and Liquid Plasmas, 2008. http://www.adastrarocket.com/ISGLP_JPSquire2008.pdf

http://www.adastrarocket.com/aarc/

Physiological problems in microgravity

"Spaceflight Might Weaken Astronauts' Immune Systems", R. Preidt, *HealthDay*, 30 Aug. 2014 https://consumer.healthday.com/kids-health-information-23/immunization-news-405/spaceflight-might-weaken-astronauts-immune-systems-690922.html

"The human immune system in space.", American Society for Biochemistry and Molecular Biology (ASBMB), *ScienceDaily*, 22 April 2013. www.sciencedaily.com/releases/2013/04/130422132504.htm

Radiations to, on, and from Mars

"Curiosity Data Shows Mars Surface Cosmic Ray Radiation Dose Rates Acceptable for Human Explorers", The Mars Society, *Mars Society Announcement*, 10 August, 2012.

"Space radiation protection: Destination Mars", M. Durante, *Life Sciences in Space Research* 1, 2–9, 2014. http://dx.doi.org/10.1016/j.lssr.2014.01.002

"Biological effects of space radiation and development of effective countermeasures", A.R. Kennedy, *Life Sciences in Space Research* 1, 10–43, 2014. http://dx.doi.org/10.1016/j.lssr.2014.02.004

"Physical basis of radiation protection in space travel", M. Durante, F.A. Cucinotta, *Rev. Mod. Phys.* 83, 1245–1281, 2011.

Mars500

About the Project "MARS—500", Institute of BioMedical Problems (IBMP), http://www.imbp.ru/Mars500/Mars500-e.html

Mars500 Blog, http://imbp-mars500.livejournal.com/

ESA's participation in Mars500, ESA website, http://www.esa.int/Our_Activities/Human_Spaceflight/Mars500

"International Symposium on the Results of the Experiments Simulating Manned Mission to Mars (MARS-500)", Abstract Book, IBMP, Moscow, 23–25 April 2012. http://mars500.imbp.ru/files/Mars500%20symposium%20-%20Abstracts%20book%20(rus+eng).pdf

NASA plan to put humans on Mars in 2033

https://www.congress.gov/bill/115th-congress/senate-bill/442/text#toc-idaef262aa-11c3-4bfe-9de5-ddf90775b6fd

https://www.nasa.gov/feature/deep-space-gateway-to-open-opportunities-for-distant-destinations

https://www.nasa.gov/sites/default/files/atoms/files/nss_chart_v23.pdf

https://futurism.com/its-official-humans-are-going-to-mars-nasa-has-unveiled-their-mission/

https://futurism.com/us-government-issues-nasa-demand-get-humans-to-mars-by-2033/

MarsOne

MarsOne, http://www.mars-one.com/

SpaceX

"Making Humans a Multiplanetary Species", E. Musk, http://www.spacex.com/mars

"SpaceX's Mars Colony Plan: By the Numbers", M. Wall, 29 Sep. 2016. http://www.space.com/34234-spacex-mars-colony-plan-by-the-numbers.html

"SpaceX unveils the Interplanetary Transport System, a spaceship and rocket to colonize Mars", S. O'Kane, 27 Sep. 2016. http://www.theverge.com/2016/9/27/13078230/spacex-mars-interplanetary-rocket-spaceship-video

Some Web Sites To Know More About

The Mars Society:
site of the international society:
  http://www.marssociety.org
Canadian chapter of *The Mars Society*:
  http://canada.marssociety.org/
Mexican chapter of *The Mars Society*:
  http://www.spaceprojects.com/Marte/
Peruvian chapter of *The Mars Society*:
  http://peru.marssociety.org/
European Mars Society Organizations:
  http://www.marssociety-europa.eu/
Belgian chapter of *The Mars Society*:
  http://www.marssocietybelgium.be/
British chapter of *The Mars Society*:
  https://marssoc.uk/
Dutch chapter of *The Mars Society*:
  http://www.marssociety.nl
French chapter of *The Mars Society*, Association Planète Mars:
  http://planete-mars.com/
Swiss chapter of *The Mars Society*:
  http://www.planete-mars-suisse.com/
German chapter of *The Mars Society*:
  http://www.marssociety.de/html/index.php
Spanish chapter of *The Mars Society*:
  http://www.marssociety.org.es/
Italian chapter of *The Mars Society*:
  https://www.facebook.com/pages/Italian-Mars-Society/106687849399082
Greek chapter of *The Mars Society*, ARES Mars Society Hellas:
  http://hellas.marssociety.org/
Polish chapter of *The Mars Society*, Mars Society Polska:
  http://www.marssociety.pl/

Bulgarian chapter of *The Mars Society*:
   https://www.facebook.com/groups/399707303378084/
Chinese chapter of *The Mars Society*:
   https://www.facebook.com/TheMarsSocietyInChina
Russian chapter of *The Mars Society*:
   https://www.facebook.com/groups/marssocietyrussia/
Indian chapter of *The Mars Society*:
   https://www.facebook.com/mars.society.india
Japanese chapter of *The Mars Society*:
   http://blog.goo.ne.jp/japanmarssociety
Australian chapter of *The Mars Society*:
   http://www.marssociety.org.au/
South-African chapter of *The Mars Society*:
   http://home.mweb.co.za/ss/ss000005/
Egyptian chapter of *The Mars Society*:
   https://www.facebook.com/The-Mars-Society-Egyptian-Chapter-
   1450635915167358/

Other interesting websites on Mars and Mars missions:

site of the Royal Belgian Observatory:
   http://planets.oma.be/MARS/home_mars_en.php
site of the Geophysics Institute of Paris (*Institut de Physique du Globe de Paris*,
   IPGP):
   http://ganymede.ipgp.jussieu.fr/GB/
site Nirgal about all aspects of Mars exploration:
   http://www.nirgal.net
site of the European Space Agency (ESA):
   http://www.esa.int/
site of ESA on the *Mars* Express and *ExoMars* mission:http://sci.esa.int/
   marsexpress
   http://exploration.esa.int/mars/46048-programme-overview/
sites of the NASA Jet Propulsion Laboratory (JPL):
   https://mars.jpl.nasa.gov/msl/
   http://www.jpl.nasa.gov/mgs
   https://mars.nasa.gov/
sites of the *Centre National d'Etudes Spatiales* (CNES, French space agency):
   http://www.cnes.fr
   https://cnes.fr/fr/media/exomars-un-radar-made-france-sur-mars-en-2020
site of the Chinese Academy of Sciences (CAS):
   http://english.cas.cn/
site of the National Space Science Center (NSSC):
   http://english.nssc.cas.cn/
site of the Institute of Geodesy and Geophysics (IGG):
   http://english.whigg.cas.cn/

site of the Technology and Engineering Center for Space Utilization (CSU):
  http://english.whigg.cas.cn/
site of the China Manned Space (CMS):
  http://en.cmse.gov.cn/

Some Other Interesting Books

"On to Mars: Colonizing a New World", R. Zubrin and F. Crossman eds., Apogee
  Books Space Series, 2002. ISBN-13: 978-1896522906.
"On to Mars 2: Exploring and Settling a New World", F. Crossman and R. Zubrin
  eds., Apogee Books Space Series, 2005. ISBN-13: 978-1894959308.
"Mars", G. Sparrow, Quercus, Revised ed. Edition, 2015. ISBN-13: 978-
  1623658564.
"Welcome to Mars: Making a Home on the Red Planet", Aldrin B., Dyson M.,
  National Geographic Children's Books, 2015. ISBN-13: 978-1426322068.
"Mars Up Close: Inside the Curiosity Mission", Kaufman M., National Geographic,
  2014. ISBN-13: 978-1426212789.
"The case for Mars—The plan to settle the red planet and why we must", R. Zubrin,
  Touchstone, New York, 1997.
"First landing", R. Zubrin, Penguin Putnam, New York, 2001.
The trilogy "Red Mars" (1993), "Green Mars" (1994), "Blue Mars" (1996), Kim
  Stanley Robinson, Bantam, New York.
"Mission to Mars, An Astronaut's Vision of Our Future in Space", Michael Collins,
  Grove Weidenfeld, New York, 1990.

Printed in the United States
By Bookmasters